FORESTS OF ASH

AN ENVIRONMENTAL HISTORY

This beautifully written and presented book tells the story of Australia's giant eucalypt, the mountain ash. It is the tallest hardwood in the world, growing to a height of 100 metres. While celebrating the steep, wet, dense eastern forests of Australia, Tom Griffiths also reveals their awesome power. Dependent on fire for its survival, the mountain ash can become a source of destruction, forcing people to confront their relationship with the bush. Death and devastation struck most shockingly on Black Friday 1939 when millions of hectares burned and 71 people died. The life cycles and fire cycles of the forests of ash span millenia. Tom Griffiths tells the ecological and social history of a unique Australian forest, and in doing so, illuminates the story of the continent as a whole.

Griffiths' narrative is supplemented by 'spotlights' written by scientists and curators from Museum Victoria about research and collections concerning the tall forests.

Tom Griffiths is one of Australia's leading environmental historians. Based at the History Program of the Research School of Social Sciences at the Australian National University, he is best known for his prize-winning *Hunters and Collectors: The Antiquarian Imagination in Australia*, published by Cambridge University Press in 1996.

FORESTS of ASH

AN ENVIRONMENTAL HISTORY

TOM GRIFFITHS

CAMBRIDGE
UNIVERSITY PRESS

PUBLISHED BY THE PRESS SYNDICATE OF THE UNIVERSITY OF CAMBRIDGE
The Pitt Building, Trumpington Street, Cambridge, United Kingdom

CAMBRIDGE UNIVERSITY PRESS
The Edinburgh Building, Cambridge CB2 2RU, UK
40 West 20th Street, New York, NY 10011–4211, USA
477 Williamstown Road, Port Melbourne, VIC 3207, Australia
Ruiz de Alarcón 13, 28014 Madrid, Spain
Dock House, The Waterfront, Cape Town 8001, South Africa

http://www.cambridge.org

First published 2001

Printed in Australia by Brown Prior Anderson

Typeface Life Roman (*Adobe*) 10/16 PT. *System* QuarkXpress® [BC]

A catalogue record for this book is available from the British Library

National Library of Australia Cataloguing in Publication data
Griffiths, Tom.
Forests of ash: an environmental history.
Bibliography.
Includes index.
ISBN 0 521 81286 0.
ISBN 0 521 01234 1 (pbk.).
1. Eucalyptus regnans – Victoria. 2. Eucalyptus regnans –
Victoria – History. 3. Forests and forestry – Victoria.
I. Title.

ISBN 0 521 81286 0 hardback
ISBN 0 521 01234 1 paperback

museum
VICTORIA

CONTENTS

PREFACE

They had not lived long enough.

These were the words of Judge Leonard Stretton to describe the people who lived and worked in the ash forests of south-eastern Australia when they were engulfed by a holocaust wildfire on 'Black Friday', 1939. The judge, who conducted an immediate Royal Commission into the causes of the fires, was not commenting on the youthfulness of the dead; he was lamenting the environmental knowledge of both victims and survivors. He was pitying the innocence of European immigrants in a land whose natural rhythms they did not yet understand. He was depicting the fragility and brevity of a human lifetime in a forest where life cycles and fire regimes had the periodicity and ferocity of centuries. He was indicting a whole society. And he was making a plea for an environmental history of the blackened forests. 'The experience of the past', reported Stretton, 'could not guide them to an understanding of what might, and did, happen.'[1]

When I was growing up in Melbourne in the 1960s, I remember my parents taking me for drives in the mountain country and talking about Black Friday. We could see the effects of Black Friday all around us. We still can today. Black Friday was Friday the 13th of January, 1939. In that week, 1.4 million hectares of Victoria burned, whole settlements were incinerated, and seventy-one people died. Sixty-nine timber mills were engulfed, 'steel girders and machinery were twisted by heat as if they had been of fine wire', and the whole state seemed to be alight.[2] Rampant flame scourged a country that considered itself civilised. Much of Victoria's mountain forest was destroyed. In the environmental history of European Victoria, there is perhaps no more significant date. I knew 1939 as the year of the great fire well before I knew it as the beginning of the Second World War. The historian Stephen J Pyne also used war to register this fire's significance: 'What the fall of Singapore was to Australian political history,' he wrote, 'Black Friday was to its environmental history.'[3] It was a loss of innocence on a cataclysmic scale.

Many of Victoria's forests of mountain ash (*Eucalyptus regnans*) were reduced to burnt ash in 1939. That fundamental irony gives these forests – the forests of ash – a perverse unity, one forged out of both their nature and their history. For fire creates as well as destroys, and the forests we know today are largely ones to which Black Friday gave birth. Black Friday is the centrepiece of this book because it illuminates the secrets of the forest's past, present and future. It was a moment in the environmental history of Australia when people had to confront – and reform – their whole relationship with the bush. Black Friday collapsed more than a hundred years of colonial history into one horrific event; it welded humanity and nature into a unique and bewildering amalgam; it reminded Australians of the enduring power of fire on their continent; and it demanded greater understanding of the past of both people and trees, not only the shared past, but the deep past stretching back millions of years. They had not lived long enough.

1

CONTINENT
OF FIRE

Australia is the driest vegetated continent – it is 'the wide, brown land' – and three-quarters of it is classified as arid. Areas with high rainfall are confined to south-western Western Australia, western Tasmania and the mainland's northern and eastern coasts. European colonists of the continent encountered what seemed like two very different kinds of vegetation, one that was antipodean, dry and alien, and the other that appeared to be northern, tropical and familiar.

The first was the melancholic, monotonous and ubiquitous 'bush', the dry, light-bleached, silver–green and blue forests dominated by the distinctive vegetation of Australia, the eucalypts, tea-trees, bottlebrushes, acacias, casuarinas, tussock and hummock grasses, banksias and grass-trees. The eucalypts in particular seemed to have the country in their grip: no other comparable area of land in the world is so completely characterised by a single genus of trees as Australia is by its gum trees.[1]

What makes the Australian flora distinctive is the adaptation (known as sclero-morphy) which produces leathery, hard, spiny, reduced leaves, a biological response to aridity, soils low in phosphates, and fire. Xerophytic (dry habitat) plants are rich in oils, making distant views more blue, fires more explosive, and tickling the nostrils of travelling, homesick Australians. These sclerophyll trees, shrubs and grasses made up what colonists quickly called 'the bush', a vegetation so dominant and ubiquitous that this amorphous phrase gestured to any uncleared area and, indeed, to the vast void beyond any Australian town or city.

Then, surprisingly, there was the 'jungle', 'scrub', 'brush' or 'fern forest'; seemingly displaced and exotic patches of wet, green, dense, enclosed forest, dangling with vines, veritable fairylands spangled into the sheltered folds of the ranges. Alfred Russel Wallace, the great British naturalist of the tropics, rejoiced in Australia's 'forests of wild luxuriance' and described how, along the Great Dividing Range of the eastern coast, 'in the midst of this apparent monotony we light occasionally on spots covered by a gigantic and exuberant growth'.[2]

Naturalist Charles Laseron described the refusal of the lichen-enshrouded beech to change for Australian conditions as 'not unlike an Englishman who persists in dressing for dinner in the wilds of Africa'.[3] What were these islands of green doing here in the wide, brown land, occupying about one per cent of the continent's area at the time of European colonisation? They were so different from the eucalypt and acacia forests, yet uncannily similar to forests in New Zealand, South America and other lands separated by vast oceans. For example, the *Nothofagus* genus (southern beech), a plant with low powers of dispersion except over land, is found today in New Guinea, New Caledonia, New Zealand, southern South America and south-eastern Australia. How did this come to be? And what could one make of the cool temperate forests where giant eucalypts (unmistakably Australian, but tall and magnificent!) mingled with the exotic rainforest trees?

It is only in recent decades that we have come to realise that the rainforest was the original forest. 'Rainforest' is a problematic word and a deceptive category. There are several different types of rainforest in Australia and they occur across the entire latitudinal spread of the continent. In spite of the name, heavy rainfall is not necessarily the distinguishing environmental determinant of rainforest; soils and fire are sometimes more critical. Some scientists therefore prefer the term 'closed forest' because a closed canopy is the major physical characteristic creating a microenvironment that is relatively constant in temperature and humidity.

In Australia, rainforest was initially seen by European settlers and scientists as the exotic element, as an un-Australian feature, alien to the image of the dry, brown continent dominated by eucalypts.[4] The British botanist Joseph Hooker, who visited Australia in 1840, noted that many species in rainforests on the east coast of Australia were characteristic of Indian, Polynesian and Malayan floras.[5] So rainforest was characterised as recent and invasive. It was described as a 'Malayan element', as 'oriental' in character, as 'Indian in its density and massiveness', as infiltrating the

Nothofagus *fossils reveal some of the deep history of the tall forests.* (Museum Victoria)

continent from the north via Torres Strait and Cape York, and perhaps also from the south across a lost land bridge from Antarctica. Charles Hedley imagined botanical Australia 'as a clear pool of water into which two muddy streams enter – a black stream from the north – a white stream from the south.'[6]

The clear pool was the pure Australian element of the eucalypts. But in the 1970s, in time to empower political activism in defence of rainforests, there was a major scientific paradigm shift, partly due to the belated acceptance of the theory of 'continental drift' or plate tectonics. It was recognised that rainforests also had an ancient Gondwanan lineage, that they had once been the dominant vegetation type across the Australian landmass, and that what remained were precious indigenous remnants, with some attributes of an Indo-Malayan intrusive flora. We now accept, in the words of noted rainforest ecologist Len Webb, 'the immense antiquity and evolutionary significance of the Australian rainforests as the progenitors of our widespread and unique flora of savanna and heath'. Rainforests were acclaimed as ancestral, as 'the most ancient Australians still surviving', as 'living fossils', and as our 'green dinosaurs'.[7]

The ocean beds that separated the continents held the key to this new narrative. Charles Darwin and Alfred Russel Wallace had already noticed another relationship between depth and time, in this case between ocean depth and known patterns of mammalian distribution. The deeper the sea, the more distantly related the fauna.[8] It was to be at the bottom of the ocean that the engine of continental movement was to be revealed. The discovery of plate tectonics was one of the great intellectual revolutions of the twentieth century and, like the Darwinian revolution of 100 years earlier, its effects and implications are still unfolding.

Scientists had earlier made guesses about Australian biogeographic affiliations on the basis of supposed land bridges between continents and the rise and fall of ocean beds. Common geological and biological signatures found on distant continents – affiliations in the fossils and living things of far-flung lands – prompted thinkers to make these courageous leaps of the imagination. And the jigsaw fit of continental coastlines had been noted from the sixteenth century. But the mutability of the world that we now accept and that underpins the new narrative of ancient Australia was inconceivable less than a human lifetime ago. Even those few people who had championed the idea of 'continental drift' from early in the twentieth century had not been able to suggest a mechanism for the movement of landmasses around the globe. Alfred Wegener (1880–1930) had been rebuffed and ridiculed when he suggested that such awesome mobility might be credited to the forces unleashed by the rotation of the earth or by gravitational forces from the sun and moon. He was himself dissatisfied with these explanations and in 1929 admitted that '[t]he Newton of drift theory has not yet appeared.'[9] Like Darwin, the explanatory power of his idea could not yet be supported by a satisfactory mechanism. Therefore, by the 1950s even sympathetic scientists acknowledged a 'marked regression away from continental drift'.[10]

But persuasive field evidence continued to overwhelm the sceptical physics. Plate tectonics, a theory which was developed in the late 1960s and early 1970s, was a revelation arising from palaeomagnetics and deep ocean exploration.[11] After the Second World War, deep ocean mapping and new geological dating techniques enabled the discovery that the seafloor is younger than the continental rocks and is being continually transformed. Where continental plates are moving apart, new sea-floor comes into being at mid-ocean ridges, and where they converge, one plate plunges down into the mantle beneath and forms an ocean trench. Inexorable seafloor spreading drives continents apart. The hard rigid masses of the continental plates therefore move on slow but powerful currents of molten rock, surfing the underlying mantle. This startling insight reveals that the earth's crust is constantly mobile and that modern, separated lands have intimate histories of the most surprising physical relationships. What the modern genetic synthesis did for evolution, plate tectonics did for earth science.[12]

Plate tectonics literally undermined Australia's history of original isolation. It revealed that the island continent only became a separate entity in the recent geological past. Most of the country's fossil history is cosmopolitan.[13] It focused our attention on a geological genesis in the southern hemisphere, followed by a relatively brief, formative journey north to lower latitudes. Just the last tenth of the 400 million year evolutionary history of the Australian flora produced 'the bush'.

So, in the beginning was the rainforest. First there were the gymnosperms, primitive seed plants such as conifers, cycads and tree ferns. Then the angiosperms, the flowering plants, evolved in the era (not more than 120 million years ago) when the southern continents began to separate. Here began those intriguing stories of co-evolution between insects and flowering plants, where plants defend themselves from insect foes but also woo insect pollinators, stories of rivalry and reward that still sustain the forests today.[14] When Australia broke away from its Gondwanan cousin, Antarctica, and 45 million years ago began its voyage north, it was covered in rain-forest dominated by the ancestors of hoop pine (*Araucaria*) and Antarctic beech (*Nothofagus*). Rafting north into warmer latitudes at seven centimetres a year, the Australian continental plate began its defining journey.

What happened along the way and over those 40 million years was such a significant 'event' that it deserves to be given a name and inscribed in our history books, and Stephen Pyne has done just that in his fire history of Australia, *Burning Bush*. He calls this transformative happening 'the Great Upheaval'.[15] During Greater Australia's lonely, latitudinal drift, the continent became embraced by fire just as its abandoned partner, Antarctica, loitering at the pole, became overwhelmed by ice. The calving of the continents itself created the circumpolar ocean current that isolated and refrigerated Antarctica. Australia journeyed to lower latitudes and became a different sort of desert.

The continent began to leach, dry and burn. Australia's ancient soils became degraded and impoverished and were hardly renewed or disturbed by glaciers or volcanoes. The land became more arid and the inland seas began to dry up. The sub-tropical high pressure systems of the 'horse latitudes' (relatively calm belts of constant high atmospheric pressure at 30° north and south) controlled the new weather. Fire became more frequent and dominant.

Under the combined assault of soil degradation, aridity and fire, the greenery of Gondwana burnished into sclerophylly. The hard-leaved sclerophyll vegetation emerged from within the rainforest to dominate and diversify, eucalypts dramatically extended their range, casuarinas succeeded araucarias, and grasses replaced ferns, moss and fungi.[16] Australia rafted north into warmer climes at a time in planetary history when the earth grew cooler, thus moderating climatic change on the continent and nurturing great biodiversity.[17] A biotic efflorescence thus accompanied geological quiescence. Rainforest retreated to 'an archipelago of refugia'.[18]

Tens of thousands of years ago, late in the Pleistocene era, Aboriginal colonists mastered the continent of fire. Fossil carbon deposits in lake beds confirm that lightning started fires long before any human habitation. Perhaps it was the smoke of fires on the horizon which first drew Aborigines to Australia. Lightning continues to cause at least one-third of south-eastern Australia's forest fires, but scholars have recently rediscovered something that Aborigines and early European settlers and explorers always knew: that fire was probably the major way in which Aborigines manipulated and changed their environment.[19]

The land that Europeans thought they had discovered was, in the words of anthropologist Sylvia Hallam, 'not as God made it. It was as the Aborigines made it.'[20] On the open plains and in the margins of the wet sclerophyll forests they kept their hunting grounds open and freshly grassed by light, regular burning. Another anthropologist, Rhys Jones, has called this 'fire-stick farming'.[21] It created open woodlands of mature, well-spaced trees. Tens of thousands of years of Aboriginal burning cultivated a squatter's dream.

To what extent did Aborigines penetrate the wet sclerophyll forests of the Great Dividing Range? Did they live there permanently, or were they seasonal visitors? Did they burn these forests? Aborigines certainly brought about great changes in some forest areas of Tasmania and Queensland. Rainforest Aborigines of northern Queensland colonised the jungle, perhaps drawing on a long cultural tradition of Asian origin. In Tasmania, the cool, temperate rainforest of Antarctic origins was a less hospitable environment. There, it was the forest that was the coloniser, taking over habitable areas and pushing people out. The Aborigines fought back with their firesticks, and it seems likely that some of the button grass plains and sedgeland on Tasmania's western coast were created by regular Aboriginal burning. It is also possible that, in some parts of Australia, Aborigines facilitated the transition from rainforest to wet sclerophyll forest through their burning. However, it is hard to tell which came first: the new landscape or the new fire regime.[22]

It is possible that in some cases Aboriginal people also preserved patches of rainforest, especially when this relict vegetation was most under pressure during the very dry and windy climate of the last glacial stage from about 25 000 to 12 000 years ago. Ecologist David Bowman speculates that 'were it not for the presence of Aborigines at the height of the last ice-age, the combined effect of fire and aridity may have been the *coup de grâce* for many species that had barely survived previous glacial cycles'.[23]

There is little doubt that their systematic fire management was an important influence on the *distribution* of rainforest in Australia, and it seems likely that Aboriginal burning led to the extinction of some fire-sensitive species. But could their firesticks have had an impact on plant evolution as well as distribution? In other words, did their landscape burning trigger or escalate the evolutionary diversification of fire-adapted non-rainforest vegetation? There is much debate about this issue and the evidence is ambiguous, but even the earliest estimates of Aboriginal occupation of the continent at over 100 000 years ago do not seem to provide enough time for such human-induced evolutionary effects to take place. The distinctive Australian sclerophyll flora evolved over millions of years. Bowman suggests that in our excitement at the discovery of how ancient the rainforest in Australia was and our eagerness to recognise (belatedly) the extent of Aboriginal environmental manipulation, it is possible that the antiquity of eucalypts has been unintentionally belittled.[24]

So, when European colonists entered the southern temperate forests of the east coast, they were late arrivals on a scene where an age-long struggle between these two great elements of the Australian flora – the rainforest and the eucalyptus forest – was still being played out.[25] Charles Laseron envisaged 'two main armies opposed to each other, the massed battalions of the jungle and the guerilla forces of the xerophytes'. Writing in 1953, he described 'a grim and relentless war of the plants, a war that has gone on ceaselessly for some 50 million years.'[26]

In tropical and monsoonal rainforests in northern Australia, the boundary between the rainforest and non-rainforest vegetation is often sharp, and the jungle rises up like a 'dark wall'.[27] As one botanist observed, the floristic differences between rainforest and eucalyptus forests can be 'so great as to suggest separate geographic and historical origins in spite of their growing side by side'.[28] But in the southern temperate mountain forests, the boundaries are blurred and rainforest can be found in the very midst of the eucalyptus forest. Lurking beneath the towering mountain ash or skulking in cool mountain gullies are sometimes found the rainforest trees, wet, green and mossy, especially the myrtle beech (*Nothofagus cunninghamii*) and southern sassafras (*Atherosperma moschatum*). The embedded nature of the temperate rainforest is a clue to the close biological and historical relationship between rainforest and eucalypts, and to the potentially transitional status of the apparently stable, 'old growth' forest of giant eucalypts that towers above.[29]

Ecological science as it first developed in the early twentieth century tended to seek out equilibrium, harmony and order in nature. Every biotic community was expected to reach a certain state of maturity or climax which was stable if left undisturbed. Often the source of disturbance was human. This definition left humans outside of nature, and nature outside of history. Historical change was an aberration rather than the norm. At the beginning of the twenty-first century, ecological science is much more attuned to the normality of disturbance, both human and natural; it is much more concerned with instability in nature, and therefore more attentive to history.[30] The breaking up of Gondwana was a disturbance, as was the coming of Aborigines, as was Black Friday. These disturbances helped make the forests of ash, this wonderfully ancient and impermanent community. If we were to eliminate all disturbances from the mountain ash forests, they would die.

Where does a eucalypt forest end and a rainforest begin? This interesting scientific question has become a pressing political one in recent decades as protesters defended mixed forests at Terania Creek in New South Wales in the 1970s and early 1980s, and as the Victorian government made promises about ending 'rainforest logging' in the 1980s.[31] The scientific definition of Australian rainforest has attracted consider-able scholarship but has become 'increasingly elusive'.[32] For example, in 1898, the

German botanist A F W Schimper defined mountain ash forests as temperate rainforest because they occurred in environments with high rainfall.[33] The existence in the forests of ash of a fully developed rainforest beneath the mature eucalypt has perplexed the classifiers of vegetation such that they have simply called this 'mixed forest'.[34] Yet other scientists have excluded the wet eucalypt forests from the category of rainforest because their vegetation is sclerophyllous. But it is more complex than that: some rainforest trees are drought tolerant and many are 'sclerophyll', in the sense of having hard leaves and low phosphorous concentration in their foliage.[35]

In his recent book on *Australian Rainforests*, David Bowman sets out to understand the patchy distribution of rainforests as scattered islands of green in a land of fire. What factors define the boundaries of rainforest and explain this patchiness? Is it rainfall, soil, drainage, light or temperature? He argues that it is fire that is the most important factor in controlling rainforest boundaries. And not just any fire, but fires of certain frequency and intensity, specific sorts and sequences of fire. Fire as an event – historical fire – has shaped the forests. The Gondwanan greenery that we know today was a product of a unique series of factors in human and natural history. The eucalypts occupying rainforest environments, however magnificent, however old their growth, may be described as 'transient fire weeds'.[36] They bring into the heart of the rainforest a deadly inheritance.

As the next chapter explores, the mountain ash is perhaps the most dramatic example of this paradox: that such impressive natural vegetation can be so prone to self-destruction.

In order to unearth a forest's secrets, then, we need to research very specific, local histories. In such accounts, it will emerge that a forest is not just any forest, but a unique community of trees, and a fire is not just any fire, but one of a particular frequency, a particular intensity, a particular range. The forest will have a history, and the fire might even have a name.

CO-EVOLUTION: A TALL FOREST STORY

KEN WALKER

Nature's secrets remain unknown unless you know where to look. Among the myriad of secrets in the tall forests is the story of the partnership or co-evolution between flowers and their pollinators. Flowers depend on a variety of measures, mainly through pollinators, to transport pollen from the male anther of one flower to the female stigma of another flower – the essence of cross-fertilisation.

Pollinators can be broadly divided into two major groups: oligoleges, those that restrict their collecting activities to a select few related groups of plants, or polyleges, those that collect pollen from a wide variety of plants. They are equally important, but I often find that the 'special' co-evolution stories revolve around the oligoleges. One such tall forest story, with a twist, involves the pollinators of *Persoonia* or Geebung.

Until 1991, the small native bee group called *Leioproctus (Filiglossa)* was known from only two specimens, one collected in 1947 in New South Wales and another collected in 1967 at Beerwah, Queensland. In the early 1990s, *Persoonia* pollination studies conducted around Sydney and bee-collecting on *Persoonia* in south east Queensland began to turn up more specimens of this strange bee group. Specimens were also collected in Victoria, which were described as a new species of *Filiglossa* on *Persoonia* from the Powelltown and Mt Baw Baw areas.

Filiglossa belongs to the primitive bee family Colletidae, which is characterised by a short and blunt glossa (tongue). Such a glossa structure is designed to collect nectar from broad, shallow cupped flowers such as occurs in the Myrtaceae. The glossa of *Filiglossa* is indeed short and blunt, however other mouthpart appendages, in particular the maxillary hairs and palps as well as the labial palps, have been extraordinarily elongated.

In 1959 the Victorian naturalist and bee expert, Tarlton Rayment, described the first known species of *Filiglossa*, *L. filamentosa*. He emphasised the excessive length of the mouthparts and suggested 'that these bees are associated with an equally remarkable flower'. Rayment's prediction of a remarkable flower has not come true, but the secret of how these bees use their extraordinary mouthparts has only recently come to light.

It seemed reasonable to assume that, since all specimens of *Filiglossa* had been collected on *Persoonia* flowers, they were oligolectic pollinators of Geebung. The basis of all science is the continued testing of hypotheses, so the *Filiglossa* pollination theory was tested using pollen washes from the bees and direct observation.

Native colletid bee
Leioproctus
(Cladocerapis)
speculiferus
(Cockerell). The
arrow points to the
short, blunt
mouthparts. Scale
line equals 2 mm.
(Museum Victoria)

Native colletid bee
Leioproctus
(Filiglossa) davisi
Maynard. The arrow
points to the
elongated
mouthparts used to
rob nectar. Scale line
equals 2 mm.
(Museum Victoria)

Examination of pollen carried by these bees showed they carried significant loads of *Persoonia* pollen, but also significant loads of pollen from other co-blooming plants. This meant that *Filiglossa* did not exclusively collect pollen from Geebung, and therefore was not oligolectic.

Direct observations revealed two more insights. *Filiglossa* is a cheat! It never gives the Geebung flower a fair chance to transport its pollen and fertilise its flowers. We found that only a bee with a body length greater than 6 mm can reach into the flower and depress one of the four tepals inside the flower to reach the nectar. *Filiglossa* is less than 6 mm, so it moves around to the bottom of the flower and pushes its elongated mouthparts between the flower tepals to reach the nectar. Since it does not enter the flower itself, it does not come in contact with the male pollen-bearing anthers. Another Colletid bee group called *Cladocerapis*, which has quite ordinary mouthparts, is the true oligolectic pollinator of *Persoonia*.

2
TALL TREES

In less than an hour's drive from suburban Melbourne, you can be embowered by ferns, ankle deep in leaf litter and dwarfed by the cathedral-like pillars of Australia's giant eucalypt, the mountain ash. Melbourne's mountain forests grow north and east of the city, occupying part of the eastern wedge of land between the Hume and Princes Highways, Victoria's two major transport arteries. They clothe the foothills and mountains of the Great Dividing Range, from Mount Disappointment in the west to Mount Baw Baw in the east, from Wandong to Walhalla, from Gembrook to Jamieson. They enclose the towns of Kinglake, Healesville, Warburton, Powelltown, Toolangi, Marysville, Rubicon and Woods Point. They are the source of Melbourne's river, the Yarra, as well as the Goulburn, La Trobe and Thomson rivers. They constitute Melbourne's water catchment area and forest playground. They once harboured many bush communities and were the state's chief source of sawn timber. In 1939, the terrible bushfire swept through them, killing people and trees. Now, as the regrowth forests reach maturity and the sawmilling industry swings its attention back from Gippsland towards Melbourne, the forests of ash loom as a conservation battleground.

The first Europeans to venture into these southern temperate forests regarded them with awe. They were different from the intensively used woodlands on Victoria's central goldfields and western plains. The special character of these steeper, wetter, denser, taller, eastern forests was expressed in the forms of the towering mountain ash, the delicate tree fern, the deciduous beech, and the theatrical and cheeky

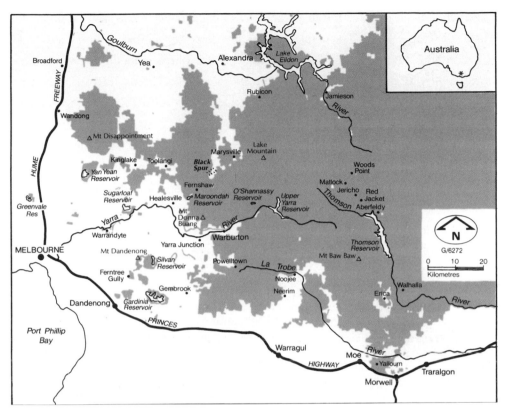

Major places and features in the Victorian forests of ash. The shaded area indicates the public land in the region, most of which is forested. (DNRE)

lyrebird. These distinctive features captured the popular imagination and have per-sisted as symbols of the region. Some colonists regarded the lushness and magic of the mountain forests as almost un-Australian; here the oft-derided eucalypt was actually impressive, and the massive ferns seemed tropical and mystical. Places were named Jordan and Jericho, a response to this sense that the forests were truly 'Eastern'. However, reverence for the forests did not stop colonists from destroying them. In fact, so vast did the forests seem, and so much did they dwarf the human figure, that restraint in their use seemed unnecessary.

In these forests are Australia's major stands of mountain ash, a tree properly described as 'the supreme expression of the genus *Eucalyptus*'.[1] Mountain ash lives in the moist mountain areas of Victoria and Tasmania and, within Victoria, occurs in three main areas: the Otway Ranges, the Strzelecki Ranges in South Gippsland (most of which was cleared for farming), and the Great Divide in the watersheds of the Yarra, La Trobe, Tanjil, Tyers and Thomson rivers on the southern slopes and of the Rubicon and Royston rivers on the north. This third area is its dominant habitat – bracketed by Mount Disappointment in the west and the Thomson River in the east. It is a tall tree with a restricted occurrence. One of its most celebrated sawmillers, Hec Ingram, fondly called it a 'parochial timber'.[2]

Less than one-twentieth of Victorian State Forest is of ash-type eucalypts: mountain ash, alpine ash (*Eucalyptus delegatensis*) and shining gum (*E. nitens*), and some forms of manna gum (*E. viminalis*). They like steep, rugged country with deep, moist, often granitic soils. The ash eucalypts occupy many of the high protected mountain slopes below the snow gums (*E. pauciflora*) and above the drier exposed foothill forests of mixed species eucalypts: peppermint (*E. radiata, E. dives*), silvertop (*E. sieberi*), and messmate stringybark (*E. obliqua*).

Whereas alpine ash is found in higher country (between 900 metres and 1360 metres), the taller mountain ash occurs principally at an elevation of between 300 and 1000 metres, and in a rainfall band of 1000–1750 millimetres a year. It dominates

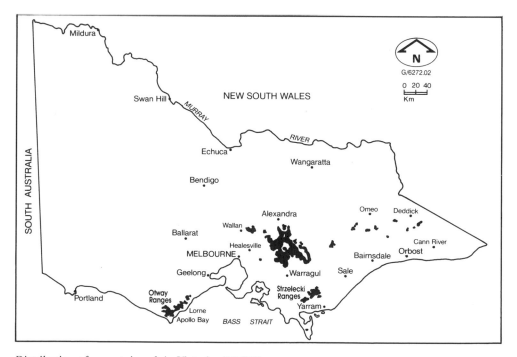

Distribution of mountain ash in Victoria. (DNRE)

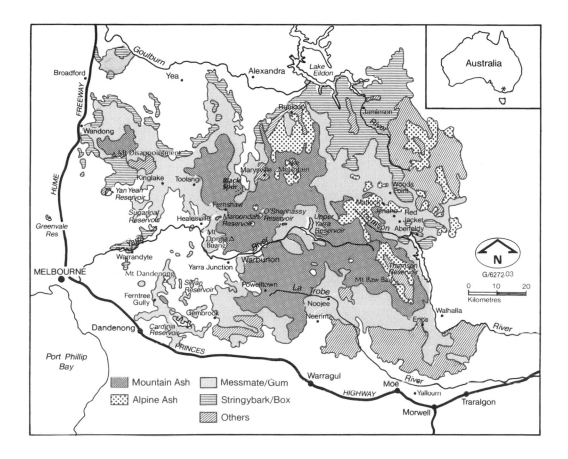

Distribution of major forest types in the region. (DNRE)

the wet sclerophyll forest and, like the alpine ash, often occurs in pure, even-aged stands. Below the high canopy of mountain ash is a second storey of trees, often silver wattle (*Acacia dealbata*) and blackwood (*A. melanoxylon*), and below them may be found shrubs such as hazel pomaderris (*Pomaderris aspera*), blanket-leaf (*Bedfordia arborescens*), musk daisy bush (*Olearia argophylla*) and various fern species. Although the drier, more open forests also pervade these ranges, it is the tall ash forests that are the focus of this history. Occasionally evidence will be drawn from other Victorian ash forests, particularly in South Gippsland, where farmers destroyed the mountain ash yet also movingly recorded its passing.

The trunk of the mountain ash is a smooth white or greenish grey and sheds long ribbons of bark. At its base clings a rough fibrous stocking. It is a very rapid grower and can reach a height of 40 metres in 20 years. It is consequently more slender than one would expect for its height, and its crown looks disproportionately small (for a forest giant) and is rather open with pendant leaves. Older trees develop buttresses at the base. Its botanical name, *Eucalyptus regnans*, means 'ruling' or 'reigning'; a scientific title that echoes its popular image as a forest monarch.

The tree was originally named as a variety of *E. amygdalina*, a Tasmanian black peppermint, but in 1870 the Victorian government botanist, Baron Ferdinand von Mueller, separated it as a distinct species. In different localities and at various times

A mountain ash 303 feet 6 inches (92.6 metres) high about to be felled by a splitter in the Narbethong district in 1888. The platform lifted the axeman above the thick buttresses at the base of the tree. (Photo by J Duncan Pierce in *The Giant Trees of Victoria*, Melbourne, 1890)

it has acquired a variety of vernacular names: white mountain ash, Victorian oak, Tasmanian oak, Australian oak, blackbutt, swamp gum, stringy gum.

It is a tree of wonder. Its secret was, and remains, its height. It is the tallest hardwood in the world. It may have been the tallest tree in the world. That belief inspired Europeans to penetrate the forest; it still does. But how could colonists accurately measure the height of a tree among other trees in densely vegetated and blustery mountain country?

The search for Victoria's (and the world's) giant trees, which took place in this forest from the mid-nineteenth century, illustrated some of the conflicting feelings about the forest. Colonists were challenged by its secrecy, felt wonder at its products and surprised themselves with a sense of loss over its destruction. It was known that softwoods, such as the North American conifers, were fast growers. They reached heights of over 90 metres, and one celebrated specimen of the Californian redwood (*Sequoia sempervirens*) stood 364 feet (111 metres) tall, but it was novel to find a hardwood with such prodigious upward growing power. Colonists were drawn into Melbourne's mountain forests by the white gleaming boles of the mountain ash, which, they said, could be seen from a distance. They rejoiced at finding 'a whole forest of ships' masts'. They pulled at the trees' hanging streamers of bark in a vain effort to gauge their height.[3]

In south-west Western Australia, the tall karri (*Eucalyptus diversicolor*) forests share many of the characteristics of the mountain ash forests of the east. The karri grows up to 90 metres in height with long, straight boles clear of branches, and needs a hot fire to release its seed and an ash bed for regeneration. In 1879, Ferdinand von Mueller compared the karri for height, girth and durability with Victoria's mountain ash.[4] But such is the chemistry of fire, rain, latitude and evolutionary history in the karri that no rainforest now mixes with this wet sclerophyll forest of the west (it was eliminated in the Pleistocene). There is enough fire to burn off any resurgent rainforest, but generally not so much as to endanger the regeneration of the tall trees.[5]

The karri and the mountain ash are unique. Elsewhere in the world there are tall, monospecific temperate forests at similar latitudes and environments – such as the redwood forests of northern California, the kauri and podocarp forests of New Zealand, and the araucarian forests of Chile – but these are all gymnosperms, whereas the karri and mountain ash are angiosperms, flowering plants, and the tallest in the world.[6]

Intercolonial and international exhibitions were occasions to boast about Victoria's tall trees. In 1866–67, thanks to the miners' tracks opening up the forest, Ferdinand von Mueller was able to publicise at the Inter-colonial Exhibition in Melbourne some astonishing (and without doubt exaggerated) figures. He claimed that there was a 480 foot (146 metre) mountain ash on the Blacks' Spur, another 420 foot (128 metre) fallen giant in the Dandenongs, and rumours of some reaching 'half a thousand feet' (152 metres) towards the sources of the Yarra and La Trobe rivers. A coach driver who took the artist Marianne North into the forest in the 1880s, told her 'they could make the Baron believe anything they liked'. But a 'woodman' also explained that he had often felled trees over 400 feet high: 'When they was down', he said. 'you could easily stump them off, and there could be no mistake about that.'[7]

In 1872 the Inspector of State Forests of Victoria, William Ferguson, assessed timber resources in 'areas that had not been penetrated by the timber splitter and the wood cutter' and reported that at lower altitudes in the Watts River catchment near Healesville he found big trees 'about ten per acre', and many of those fallen measured 350 feet (106 metres) in length. 'In one instance', he reported to Clement Hodgkinson, the Assistant Commissioner of State Forests, 'I measured with the tape line one huge specimen that lay prostrate across a tributary of the Watts and found it to be 435 feet [132 metres] from its roots to the top of its trunk.' Ferguson considered that 'before it fell it must have been more than 500 feet high' because it had been much burnt by fire.[8]

The *Illustrated Australian News* of 10 June 1878 featured a tree supposedly recorded at Fernshaw and said to be 380 feet (116 metres) to its first branches and 430 feet (131 metres) to the top. A huge stump, 80 feet (24 metres) in circumference, was taken for display at the Melbourne International Exhibition of 1880. When the famous English writer, J A Froude visited the Australian colonies in 1884–85, he was taken to see the giant trees on the Blacks' Spur, which he was told were 'the largest as yet known to exist anywhere, higher by a hundred feet than the great conifers in the Yosemite valley'. As he travelled into the Upper Yarra, columns of smoke were rising in half a dozen directions from the mountains, which Froude guessed to be 'from forest bush-fires; either lighted on purpose to clear the ground, or the careless work of wood-cutters or wandering natives', and everywhere he spied 'the genius of destruction', as evidenced by burnt and ringbarked trees. Yet, on the Blacks'

A mountain ash featured in The Illustrated Australian News *of 1878, said to be 430 feet high (131 metres).*

Spur he rejoiced that 'one drives as through the aisles of an immeasurable cathedral' beneath trees with an average height of 350 to 400 feet, 'and one was measured which reached 460' (140 metres). 'It was something', he exclaimed, 'to have seen the biggest trees in the world, and to be able, in California, to affect disdain of Yosemite'. Here, in southern Australia, he was delighted to find a part of the New World that was both big and British.[9]

As time passed, as measurements became more precise and – a sobering thought – as timber splitters claimed more choice victims, the reported heights dropped. At the time of the Centennial International Exhibition held in Melbourne in 1888, a reward of £20 was offered to anyone who could locate a tree 400 feet (122 metres) in height; there was to be an extra £3 for every five feet (1.5 metres) above that. No such tree was found, but several over 300 feet (91 metres) were identified, and the tallest – on a spur of Mount Baw Baw – was measured at 326 feet one inch (99.4

metres). The tallest living Australian tree ever measured by a qualified surveyor was one known as the Thorpdale Tree, in South Gippsland. It was measured by George Cornthwaite with a 5 inch (127 millimetre) theodolite in 1881, and promptly chopped down. It was 375 feet (114 metres) high. A concrete pole topped by a sign declaring it the site of the 'The World's Tallest Tree' now marks the spot.[10]

Claims were frequently made that somewhere in the forest there were indeed the world's tallest trees. Perhaps they still remained to be found. Perhaps they had already been cut down by paling splitters. Stories abound in the bush of 400-footers logged years ago. As naturalist Crosbie Morrison said in 1944, 'It's rather like fishing yarns – there's always the doubt about 'the one

A tall pole in a cleared landscape in South Gippsland marks the site of the 'The World's Tallest Tree', which was measured at 375 feet (114 metres) high in 1881, and promptly chopped down. (Photo: Meredith Fletcher)

that got away', though, in the case of the trees, it's the one that was cut down in the "eighties"'. Even A V Galbraith, a chairman of the Forests Commission of Victoria (1927–49), stated in 1937 that 'Mountain Ash can make serious claims to being the world's highest tree'. His successor as chairman, Fenton Gerraty (1949–56), himself measured a fallen tree found near Noojee after the 1939 fires that was 338 feet (103 metres) high, even with its top tantalisingly broken off.[11]

Tall and ancient trees earned recognition not just as members of a species but as individuals. They became visited, studied, named and beloved. This level of distinction was a sign of change in the forest. Old trees became valued as historic artefacts as much as botanical specimens. The age of gum trees was a puzzle to colonial scientists who soon found the counting of tree rings to be an unreliable guide to years of life in a land where seasons were rarely annual. Mature eucalypts, especially those of the size and grandeur of the mountain ash, were assumed to be much older than they were, in the realm of thousands of years. Ferdinand von Mueller considered them to be 'of a far more venerable age than is generally supposed', speculating that some specimens 'stood already in youthful elegance, while yet the diprotodon was roaming over the forest ridges encircling Port Phillip Bay'.[12] As late as the 1940s, naturalist A G Campbell guessed that giant specimens of mountain ash were '2400 years of age or about as old as man's written history'.[13]

Novelist Marcus Clarke found that ancient eucalypts evoked 'thoughts of the vanished past which saw them bud and blossom'. They were, he wrote, 'fit emblems of the departed grandeur of the wilderness'.[14] The Victorian landscape photographer, Nicholas Caire, photographed, measured and named many outstanding specimens of mountain ash in the late nineteenth century. He observed that nearly all the giant trees he had located were decaying and had lost their tops, and guessed an age of over 3000 years for some. He called these denizens of the 'aboriginal' or 'primeval' forest 'the oldest inhabitants in the land', and regarded them as yet another passing race making way for the settlers. He gave them regal names not unlike those his society bestowed on Aborigines identified as 'the last of the tribe': King Edward VII, Big Ben, Uncle Sam. And he photographed them for posterity, lest future generations, bereft of giant trees, doubted that they had ever existed.[15]

King Edward VII in the Cumberland Valley had a girth (generally measured at breast height) of 21.3 metres and was destroyed by fire before 1920. Big Ben on the Blacks' Spur, with a girth of 17.4 metres, was caught in the 1902 fire. Caire's epitaph

At the turn of the century, Nicholas Caire named and photographed many of Victoria's
remaining giant trees. This one is King Edward VII with a girth of 70 feet, pictured in 1904.
(National Library of Australia)

for this giant was: 'He probably was a sapling when the people of England were semi-barbaric.'[16] Uncle Sam, beyond Fernshaw, measured 14 metres in girth and reached 76 metres upward, even with a broken top. Caire lovingly measured its shrinkage as it aged. Cobb & Co. coaches passed it every day and it became a renowned artefact. In 1933 Harold Furmston discovered a magnificent mountain ash with a girth of 19.5 metres on Mount Monda. The Healesville Shire President led an excursion to the tree which became a popular site of pilgrimage for hikers. 'Furmston's Tree' was nominated a National Monument by a sub-committee of the Field Naturalists Club of Victoria in 1945. It collapsed at the end of the century.

Today the finest stands of mountain ash are to be found in Melbourne's water catchment areas. At Wallaby Creek, Watts River and O'Shannassy there are mature stands of ash that include trees over 300 years old and nearly 90 metres in height. 'Mr Jessop' is such a one, named by the botanist David Ashton after the chairman of the Board of Works (1940–55), for his role in encouraging old-growth forest research. At about 85 metres in height and 10 metres in girth, 'Mr Jessop' is one of the tallest living trees in Victoria.

A more accessible giant stands in the headwaters of the Little Ada River north-east of Powelltown. The 'Ada Tree' was located and named by Werner Marschalek and his brother Joseph in 1986. Probably more than 300 years old, the tree stands over 76 metres high and its girth is an impressive 15 metres. A Friends Group has constructed a delightful walking track which weaves 1.6 kilometres through cool temperate rainforest to a protective boardwalk at the base of the big tree. Other specimens of mature mountain ash can be visited in the Cumberland Valley near Marysville, where a 'Sample Acre of Tall Trees' was set aside in the 1920s. Before fierce winds and old age did their damage, this reserve boasted 27 mountain ash with an average height of 81 metres including a titan, the Big Tree, reaching 301 feet (92 metres).[17]

We will never know now whether they were the world's tallest. The systematic search for the big trees came twenty years too late. When the forests of ash were first penetrated by Europeans in the middle of the nineteenth century, tall trees, trees of giant girth and decaying or fallen trees were common. Even in the early twentieth century some keen bushwalkers did not take tents but relied on finding large hollow logs to sleep in. Today, mature trees and large hollow logs are rare. Perhaps a quarter of the total mountain ash area in Victoria has been cleared for dairying and agriculture, another quarter has been destroyed by frequent fires and replaced by scrub and bracken, and a large proportion of the remaining half is regrowth forest of greater density.[18]

Ash-type species are different from most other eucalypts in their means of re-generation. Although they produce epicormic shoots from their branches after fire, drought or insect damage, they do not develop ligno-tubers under the ground from which they can renew themselves, and mountain ash and alpine ash rarely coppice (grow new shoots from a cut stump). For their survival, therefore, they are unusually dependent on their seed supply. Mountain ash dies out unless fire periodically sweeps the forest, for it is principally fire that releases the seed from the tree's hard capsules. However, the tree is also unusually sensitive to fire. Its bark is thin, and mature trees are easily killed by fire. Furthermore, if a second fire comes before the regrowth has developed its own viable seed, a whole forest can be wiped out. The name 'ash', however, is not expressive of colonial ecological insight but of immigrant nostalgia. To British settlers, the timber bore a superficial and hopeful resemblance to English Ash.

The well-named botanist, David Ashton, has devoted his life to the study of the ecology of the mountain ash. He has elucidated the apparently paradoxical relationship between ash and fire until it seems instead 'a miracle of timing'.[19] Mountain ash generally occurs in even-aged stands. That, and the persistence of soil charcoal, are evidence of past catastrophic fires. Mountain ash is very sensitive to light surface fires, but seeds prolifically in intense crown fires. In fact, it possesses features that seem to promote such fires: a heavy fall of inflammable leaf litter (two or three times that of other eucalypts), particularly in dry seasons, hanging streamers of bark that take the flames up to the canopy and become firebrands hurled by the wind in advance of the flame, and open crowns whose pendulous foliage encourages updrafts.

And how do these precious seeds survive the intense heat that they indubitably need? Ashton suggests that perhaps it is the very flammability of the crown that protects the seed in its capsule – just long enough. In the crown of the tree ahead of the fire front, the heat is brief and explosive and, some observers say, is followed by cool updrafts of air before the arrival of the surface fire. This fragile and complex circumstance certainly works. At Noojee after the 1939 fires, forester A H Beetham found that nearly 2.5 million seedlings per hectare of mountain ash germinated.[20]

The forests of ash are adapted to and a product of intense crown fires that occurred once every few hundred years. Mountain ash grows in high rainfall areas on southern slopes protected from desiccating northerly winds. Fire generally intruded only after several years of drought and on very hot windy days, but when it did intrude it raged high and far. If these conditions never erupted and no fire swept the forest during the 400-year lifetime of ash, the ash forest died out. Sometimes beneath the tall mountain-ash canopy a fully developed rainforest of myrtle beech and southern sassafras bides its time, ready to become dominant in the long absence of fire, ready to re-assert its ancestral priority. If fire swept the forest too frequently – at intervals of less than 15 or 20 years, as has happened since European settlement – the young ash did not have time to produce seed and again the ash forest died, replaced this time by bracken and scrub. The very existence of mature ash forests, such as those found by Europeans in the mid-nineteenth century, is testimony to a fire regime of very occasional widespread but intense conflagrations.[21]

Melbourne's forests of ash fall within the boundaries of two Aboriginal tribes: the Woi wurrung and the Daung wurrung, who with other clans formed the 'Kulin nation'. Early white accounts of the area confirm that the Woi wurrung and the Daung wurrung did use this forest, if only seasonally. Surveyor-General Robert Hoddle followed the Yarra to its source in 1845 and marked on his map the site of a 'Black Aboriginal old encampment' near the present Upper Yarra dam picnic area. He also recorded a pathway along the ridge top overlooking Contention Gully. Hoddle reported that he 'met with no blacks, but the barking of trees show that one or two have passed a considerable period back'.[22]

Robert Brough Smyth, mining engineer and author of *The Aborigines of Victoria* (1878), recorded that Aborigines made seasonal visits to 'the glens and ravines on both sides of the chain [Great Dividing Range], but they did not live there'. Their trips, he said, were made with the specific purpose of 'obtaining woods suitable for making weapons, feathers for ornament, birds and beasts for food, and for the tree fern, the heart of which is good to eat, and for other vegetable productions.'[23]

Several accounts also survive of groups of Aboriginal men travelling to the Dandenong Ranges and possibly further for lyrebird tails. The Aboriginal Protector in the Westernport area, William Thomas, noted that July was the most popular time of the year for such visits as the lyrebirds were then 'in their prime'.[24] When botanist Daniel Bunce walked with several Aborigines into the Dandenong Ranges in 1839, he saw lyrebird scratchings everywhere and was impressed with the 'great beauty' of the tails that his friends brought to him. Their diet consisted of possum and kangaroo, occasionally echidna and wombat, and 'the heart or crown of the fern tree, slightly roasted'. He found it 'an acceptable dish, the taste of which reminded me of the flavour of the cocoa-nut'.[25]

Later European forays into the forest found evidence of Aboriginal occupation, but not the people. In the early 1930s, R A Keble of the National Museum of Victoria visited the Upper Yarra Valley and, while searching for graptolites, located grinding grooves and an edge-ground hatchet head on the Woods Point Road.[26] In the late nineteenth century, Walhalla schoolteacher and naturalist Henry Tisdall was shown stone axes exposed by farmers clearing the forest near the Tyers River: 'From the number of axes found it may be concluded that the tribe was of some importance'.[27] In the 1870s and 1880s land selectors penetrated and cleared much of the tall forests of south Gippsland and found stone axes aplenty, 'far too numerous to have been dropped by occasional visitors', according to T J Coverdale, one of the settlers. He was convinced that Aborigines had occupied 'the scrub' (as the settlers called the tall timber) and that they had periodically fired it. He certainly held Aborigines responsible for 'Black Thursday', the great fire that engulfed Victoria in 1851:

> The blacks were no doubt the originators of the fires, whether accidentally or otherwise. Perhaps the strategy of some sable Napoleon during the operations of a Summer campaign may have demanded the burning of portions of the scrub to embarrass the enemy, or to cover a masterly retreat; and so a conflagration would be started.

Coverdale remembers with puzzlement the rare bare patches of scrub along ridgetops in South Gippsland. Were they remnants of a once open forest? He thought not. Were they the result of some deficiency in the soil? Or were they a consequence of regular Aboriginal burning of favoured pathways?[28]

Understanding of, and respect for, traditional Aboriginal burning practices have increased greatly in recent decades, particularly in northern and central Australia where exciting work is being done to bring together Aboriginal natural ecologists and university trained scientific ecologists to discuss ways of promoting biodiversity.[29] But the grass-borne fires of the savanna lands and their continuous traditions of Aboriginal burning contrast with the dangerous crown fires of the southern tall forests and their less certain depth of human history and management.

There is a frustrating lack of knowledge about Aboriginal use of the forests of ash before Europeans arrived in Australia. Archaeologists have focused on more accessible and closely inhabited coastal areas, river valleys and open plains, and systematic investigations have only recently penetrated the mountain forests. Even then, attention was inevitably given to areas where some disturbance had already exposed the ground, such as along fire trails, logging roads or around old log dumps.[30] The mountain forest, with its poor ground visibility, deep leaf litter and rugged terrain, makes site surveys very difficult. Surface sites could easily exist and never be found.

The first systematic archaeological study of the Upper Yarra Valley and Dandenong Ranges region was undertaken as late as 1987 by Hilary du Cros, and the first survey of the Big River and upper Goulburn/Black River valleys was completed in 1990 by Cath Upcher. Du Cros found four artefact scatters – the residue from tool-making – and several isolated artefacts around the margins of alpine swamps on the Toorongo Plateau, as well as on a track at Paradise Plains, at Old Warburton and near Powelltown. Du Cros also found stone (chert) in the mountains that had been brought from the Mt William hatchet stone quarry, a significant ceremonial exchange centre near Lancefield in central Victoria. Upcher located eight sites of artefact scatters or isolated artefacts along a terrace to the south of the Big River. Most of the flakes were of silcrete, a stone that does not occur naturally in the Big River Valley. Archaeological surveys of forested areas elsewhere in Australia have found similar evidence of Aboriginal occupation along ridges, in saddles, on flat terraces above rivers and on gentle mountain slopes.[31]

The forests of ash were probably not permanently occupied by Aborigines but visited seasonally – briefly in winter for lyrebirds and principally in the summer by hunting and foraging parties. Campsites would have been used along the length of major rivers that penetrated the highlands, such as the Yarra and Big Rivers. Aborigines probably did burn the margins of the wet sclerophyll forests and may have maintained clearings and pathways in forest areas (particularly along ridgetops), but it seems unlikely that they systematically burnt the mountain forests in the way they did the drier forests and plains.

In 1890, the eminent explorer, naturalist and anthropologist, Alfred Howitt, presented a paper to the Royal Society of Victoria on 'The Eucalypts of Gippsland' in which he offered an intriguing environmental history of Aboriginal dispossession and European settlement. When the first European pastoralists drove their flocks and herds down the mountains from New South Wales into the pastures of Gippsland they initiated an unexpected environmental revolution, argued Howitt. Before the arrival of Europeans, Aboriginal people had controlled plants and insects with their annual fires, and grazing marsupials with their hunting, and this had kept the forests healthy, open and well-grassed. The new settlers, keen to protect their homes, fences and animals, suppressed fires and overgrazed the grass, which meant that '[y]oung seedlings had now a chance of life, and a severe check was removed from insect pests'. In some places, the forests thickened and encroached upon previously open areas; in others, whole forests died as they came under attack from insect plagues. Encroachment and dieback: Howitt provided an early, species-specific account of two phenomena that still challenge land managers today.[32]

In the mountains north of Maffra and just east of the forests of ash, Howitt
reported that the increase in vegetation was very marked: 'These mountains were, as
a whole, according to accounts given me by surviving aborigines, much more open
than they are now.' He also observed that, in the great ash forest of South Gippsland,
'there are substantially only two generations of trees'. One was of a few very large old
trees, and the other was of numerous young trees originating no earlier, he thought,
than the Black Thursday fire of 1851. He considered that the number of Aboriginal
stone tomahawks found during the process of clearing this country, now covered by
a dense scrub, suggested that it had formerly been open forest. Howitt, renowned for
his sympathetic approach to Aboriginal society and his close association with Kurnai
people, concluded:

> I might go on giving many more instances of this growth of the Eucalytpus
> forests within the last quarter of a century, but those I have given will serve to
> show how widespread this re-foresting of the country has been since the time
> when the white man appeared in Gippsland, and dispossessed the aboriginal
> occupiers, to whom we owe more than is generally surmised for having
> unintentionally prepared it, by their annual burnings, for our occupation.[33]

Howitt's scientific audience in 1890 was intrigued but distracted. President of
the Royal Society, the professor of engineering W C Kernot, admitted that '[t]he
re-foresting of parts of Gippsland was a very interesting fact, and one he confessed
he had never heard or dreamt of before.' This was 'an opposite action', he said, to the
one of which they were more conscious: 'that the country had been denuded of its
forests'. Kernot and others then directed the discussion away from the human history
of forests to the endlessly fascinating subject of tall trees, how high they might once
have been, and whether you could still find them. Howitt's brief, passing mention of
a mountain ash reported to be 415 feet tall effectively robbed him of close attention
to his broader thesis, for members were 'very glad indeed that the question of the
height of the eucalypts had been mentioned', and away they went. Ferdinand von
Mueller was among them and, while congratulating Howitt on his timely attention to
the question of the distribution of eucalypts, he took the opportunity to moderate
some of his earlier claims about the height of the ash and defended 'nearly 400 feet'
as a possible maximum.[34]

Tall trees and rainforests continued to hijack the conservation debate a century later. With the energy and observation of a latter-day Howitt, the farmer, poet and historian Eric Rolls wrote an important book in 1981 that again drew the attention of Australians to the history of our drier forests. This classic study, *A Million Wild Acres*, was set in the Pilliga Forest in northern New South Wales, much of which grows on what were once good pastures.[35] The book confronted and provoked Australians with the idea that in many areas of the country, landscapes that had once been grassy and open are now densely vegetated, in many places there might be more trees in Australia now than at the time of European settlement, and that some forests – which we so readily and romantically see as primeval – could actually be the creation of a recent act of settlement.

How many trees make a forest? he asked. 'Australia was not a timbered land that has been cleared', argues Rolls.[36] In much of Australia, Aborigines kept the forests open with their light and regular burning. The prolific germination that always follows fire in Australia was kept in check by the plentiful wallabies, possums, bandicoots and rat kangaroos, which ate the seedlings. In the absence of Aborigines and small marsupials, forests thickened and extended their range, and long-lived colonists remembered playing and working where trees and scrub later grew. Without Aboriginal fire management, occasional and intense wildfires erupted, a product of European occupation. Today's forests, Rolls reminds us are mostly not remnants of a primeval jungle: 'they do not display the past as it was, they have concentrated it.'[37] They are different and new; they are exaggerated communities of plants and animals; they are especially vulnerable.[38]

When Rolls was writing *A Million Wild Acres*, the conservation battlegrounds in Australia were the rainforests, most notably at Terania Creek in northern New South Wales in 1979.[39] As Rolls acknowledges in his final chapter, woodchipping was also an issue and had become shorthand for indiscriminate forest clearing and exploitation. Rolls considered it a necessary industry committed to unnecessary destruction.[40]

So his book was written in the midst of those campaigns, when forests were depicted as timeless and primeval, and human disturbance seemed only to mean the destruction of trees. Just as Howitt sought to balance the colonial fascination with tall trees and loss by offering a more complex environmental history, Rolls engaged with his society's concern over rainforest logging by writing a detailed regional study showing that forests could also be the creation of settlement, and that these 'phoenix forests' were differently vulnerable. And like Howitt, Rolls gave voice to a myriad of local observers and saw system and integrity in a scattered range of puzzling personal accounts.[41]

The magnificent mountain ash forests that exist today are also products of history with a distinctive ecology, and they resist some of the generalisations about rain-forests on the one hand, and dry forests on the other. They have little resistance to fire. Their regeneration is precarious. They were not burnt lightly and regularly by Aborigines. Holocaust fires are endemic. Vast areas of mountain ash were cleared by settlers, and other areas were denuded by repeated fire.

When Europeans first entered the tall forests, their testimony varied as to its density because of the variety and localised nature of fire regimes. In places they recalled riding a horse through the scant undergrowth; in others they had to fight through thick scrub to make a pathway.[42] Where they did find an open forest floor, it was not due to regular burning but to the long absence of fire. The mature mountain ash forests excited them, and the height and girth of the giants evoked a history beyond people. *Eucalyptus regnans* was not quite as tall or as individually old as colonists thought, but it has an intriguing human history.

ASH AND ANTS

ALAN L YEN

Physically, mountain ash trees and ants sit at the extreme ends of the scale. Ants counteract their small size by sheer numbers and activity, and by their ability to travel from the ground level into the canopy of the trees.

The canopy is home to a variety of plant-feeding insects, such as stick insects, psyllid bugs and leaf-chewing beetles, predatory groups such as spiders, and pollinators such as native bees. One of the larger bodied insects in the canopy is the tree cricket which shelters in curled up strips of bark during the day, and emerges at night to catch and feed on insects and spiders. These tree crickets form an important part of the diet of Leadbeater's Possum.

The invertebrate fauna on the ground lives in the litter, soil or in rotting wood. Some of the groups involved in the decomposition of leaves, bark and fine, woody debris include earthworms, amphipods and isopods (both crustaceans), millipedes, and the larvae of flies and moths. There are at least two species of earthworms, including the relatively large *Megascolex dorsalis* (around 30–38 cm in body length). When the leaf litter or soil is disturbed, there is a mass of jumping amphipods, appropriately named 'random hysterias', where they occur in densities of up to 400 per square metre. These amphipods are eaten by lyrebirds, which turn over a considerable amount of litter and top soil when searching for food. More secretive in their behaviour are the termites; they are found inside rotting wood, and are most noticeable when the winged forms make their nuptial flight.

While not as obvious as amphipods, the ants go about the forest involved in a number of different ecological roles. Some ants are active hunters, others are scavengers, some tend sap-sucking insects or the larvae of leaf-feeding caterpillars, and some harvest plant seeds. It is estimated that there are 5–6 million ants per hectare.

Unlike some plant species that have a seed bank in the soil, eucalypts retain their seed in the capsules in the canopy and release them for germination. Ants are among the major destroyers of mountain ash seeds. At least four species of ants harvest a large proportion of the seeds dropped by mountain ash trees – it is estimated that *Prolasius pallidus*, *P. brunneus*, *P. flavicornis* and *Chelaner leae* remove 60 per cent of the seeds that fall during the year. As David Ashton's work has shown, it is only after a periodic wildfire, when there is an enormous seed drop and the environmental

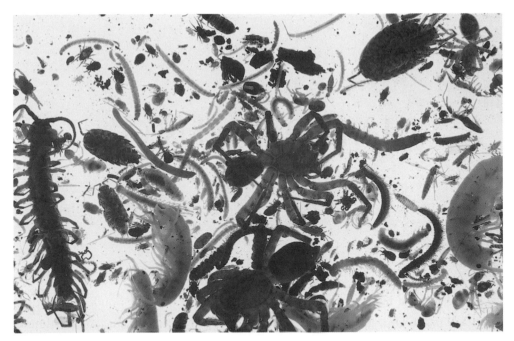

A sample of the high biodiversity of invertebrates that live in the plant litter on the ground of mountain ash forests. (Museum Victoria)

conditions are conducive to seed germination, that a larger proportion of seeds escape ant predation.

Fire reduces the diversity of invertebrate groups in mountain ash forests considerably, although most have recolonised after five years. The number of ant species found after a wildfire is similar to that before the fire, but the composition can be different, and in one case F G Neumann found that the seed harvesting *Prolasius pallidus* occurred in larger numbers for 12 months after the fire. The invertebrate groups that are significantly affected by wildfire are the amphipods and isopods. Amphipods are absent from the forest for at least 12–24 months after the fire, and we have yet to learn about the recolonisation of forests by amphipods after fire.

While mountain ash trees physically dominate the forest, the smaller animals have significant ecological roles. Together they are involved in complex ecological functions that we are still trying to unravel.

3

'IMPROVING'

W K Hancock, in his pioneer environmental history of the New South Wales Alps, *Discovering Monaro*, has observed that the word 'improvement' was an early immigrant to colonial Australia. In its usage, he wrote, 'we hear intonations of nostalgia: improvement of "the new country", it seems, means doing everything that a man can to make it look like "the old country"'.[1] 'Improvement' was nostalgic; it was dismissive of indigenous environmental systems; it was aggressive as well as progressive. This nostalgia also had social and moral dimensions: 'improvement' very often meant the settlement of an idealised yeomanry, of self-sufficient family freeholders. Colonists wanted to see small fields carefully tilled. In this sense 'improvement' was as much about what the land could do to people as what they could do to the land. 'We are going to grow tall men where there are tall trees' was the political catchcry. 'Improvement' was not just an attitude. It became a moral and legislative measure of possession, even a social strategy.

Europeans colonised Port Phillip for its grasslands. They wanted open pasture and permanent freshwater for their sheep and cattle, so they spread out across the flat open spaces to the south-east, north and west of Melbourne. The basalt plains became their kingdom. They were not attracted to the rugged, thickly vegetated mountain country in the east, the mountains that were the source of the Yarra.

However, some squatters did follow the broad river valleys and occupy the accessible alluvial flats and floodplains of the Yarra River. An overlander from Sydney, John Gardiner, came across rich pastures in the Croydon–Mooroolbark area when seeking lost cattle in 1837. In the late 1830s, William, Donald and James Ryrie set out from Braidwood on the Southern Tablelands of New South Wales. With their large herd of cattle they settled on the river flats near the future site of Lilydale and called their property by the name they thought the Aborigines gave it, Yering. The grapes they planted were inherited in 1850 by the de Castella family, which later established a local wine industry. To the south, the Reverend James Clow took possession of Corhanwarrabul Run, which included land at the foot of the Dandenong Ranges.

Gulf Station, which survives today as a historic homestead on the eastern outskirts of Melbourne, was farmed by the Bell family for 100 years from the 1850s. The Bells, farmers from Scotland, arrived in 1839 on the first ship bringing assisted immigrants to the colony of Port Phillip. They had been recruited to supply farming skills to the new colony, to grow grain for the expanding population of Melbourne. They very quickly established themselves as farmers in the Eltham–Kangaroo Ground district and then, during the 1850s, at Yarra Flats. The homesick Scots renamed that broad floodplain Yarra Glen.[2]

By the end of the 1840s squatters embraced, but did not intrude upon, the ash forests. They occupied most of the plains and open valleys, but made no claim on the higher, densely forested terrain. A few, perhaps, followed the watercourses into the mountains and grazed the grassy river flats, but squatters were secretive about new discoveries.

Although few in number, squatters had a significant ecological impact. Squatting runs were often larger than the hunting territories of the Aboriginal bands on which they were superimposed.[3] The lonely names of their runs printed across the vast spaces of early pastoral maps can be misleading. Each squatter was accompanied by thousands of cattle and sheep, whose hard hooves compacted the delicate vegetation and light soil, muddied and eroded precious watercourses, and introduced and spread new grasses and seeds. The displacement of Aboriginal people and the dingo created the conditions for an explosion in the kangaroo population. South of the forests, at Western Port, kangaroos were observed to be 'so plentiful they resembled flocks of sheep', and their ubiquity was partly because of the sheep: their browsing exposed the shorter grasses.[4]

The Land Acts of the 1860s were designed to break the dominance of the squatters and to 'unlock the lands', as the persistent political catchcry put it. The attempt to settle 'yeomen farmers' dominated the alienation of land in Victoria for a century. In forested country, trees rather than squatters were the selector's enemy. Trees were initially regarded as an impediment rather than a resource. Ringbarking was the supreme expression of this attitude. It killed the tree with minimum labour and let in the light to encourage grass growth. Some said it dried and sweetened the soil. Others argued for a balance 'between area under forests and land under cultivation' and were concerned that such clearing, which was 'something enormous', would lead to 'unwelcome climatic changes at no very distant date'.[5]

Turn-of-the-century photographs of settlement in the Strzelecki Ranges and the foothills of the Great Divide reveal that people carried out their lives amid bleak panoramas of stark white trunks. Dorothea Mackellar wrote in her poem 'My Country' of these ringbarked forests 'all tragic to the moon'. One Gippsland settler used two giant stumps as gateposts, exclamation marks of occupation. Some even lived in them. Nicholas Penny arrived in Fumina (east of Melbourne) in 1905, and one of his first tasks was ringbarking trees, one of which was to become his home. He lost his tent and possessions in the 1906 fire. The tree was burnt inside but remained quite solid; so he moved in. It was 6 metres across at floor level, and he lived there for nine months with his wife, Eva, and their three children.[6] At Neerim a settler used a giant tree as his capacious three-storied dwelling, and in South Gippsland a teamster stabled his twelve bullocks in one.[7]

The tussle over the proper balance of cleared and forested land wrote its history on the landscape. In the late 1880s and early 1890s, forester J. Blackburn reported on Victorian timber reserves and state forests, often declaring that 'I fear that a mistake has been made both by the Department in offering this country for selection and by selectors in taking it up. The timber should have been preserved as in many places ... it is magnificent.'[8] The Secretary of the Forests Commission of Victoria put it this way in 1924: '[I]n most parts of the State today burnt and ring-barked skeletons bear mute witness where fine timber has been destroyed, but settlement has not been able to make good'.[9]

Most Victorian public land is forested, and most forests are on public land. This is because the first signature of projected ownership was a clearing. The story of the definition of public land was partly the realisation that the forest resource was finite. Much later it was the recognition that cleared land was often vulnerable land.

The reservation of timber as a resource was frequently a cause of bitter contention. Forests limited and constrained settlement, and settlement was regarded as the highest public benefit of all and the true measure of progress. The settler or 'pioneer' was an heroic figure. However, even the pioneer's heroic adversaries were partly self-inflicted. Burning off, and the ringbarking and wasteful felling of trees, heightened the risk and severity of fires. Clearing hill-slopes heightened flood danger in winter, increased the risk of drought during summer, and caused massive soil erosion.[10]

In forest country, clearing and improving went together. A witness to the Royal Commission inquiring into the 1939 fires explained:

> After the gold rush was over, the white man had to make use of the land and he had to get rid of the timber. He slaved, toiled and burned to get rid of it … The children and grandchildren of these men have grown up with minds opposed to timber.[11]

Settlers were often overwhelmed and frustrated by the quantity of bush timber, and clearing it was seen not only as a private necessity but also as a public good. In 1855 the visiting English writer William Howitt erroneously referred to Australia as 'one huge, reclaimed forest' which he believed was unhealthy due to the mass of uncleared vegetable matter it generated. 'All these evils', he wrote, 'the axe and the plough, and the fire of settlers, will gradually and eventually remove'.[12] These were indeed the instruments of destruction.

One selector in eastern Victoria estimated that, in the first five to ten years on the land, nine-tenths of the labour was devoted to axe-work.[13] This fight with the forest assumed theatrical dimensions in South Gippsland, where each summer neighbours gathered to watch the giant burns that, they hoped, would turn last year's fallen and ringbarked forest into this year's clearing. These self-conscious pioneers memorialised their defeat of 'the Great Forest' in a remarkable book called *The Land of the Lyre Bird*, which they published in a cleared landscape in 1920. The volume brought together the reminiscences of over fifty selectors in the district, men and women; it is heroic and elegiac, and laments the forest's passing even as it celebrates it. For these settlers, the lyrebird became the symbol of the mystery and beauty which, in their victory over the scrub, they had lost.

Many of the settlers could hardly believe that they had themselves cleared vegetation of such magnificence, 'one vast forest as dark as night' where the 'trees towered up till their tops seemed lost in space', and where 'there reigned a strange and oppressive stillness, broken only by the notes of the lyre bird or grunt of the monkey-bear [koala]'.[14] They needed to establish pastures as quickly and cheaply as possible. Small trees were chopped, undergrowth was slashed, and sometimes large trees were felled so as to demolish smaller timber that had been previously 'nicked', thereby creating 'a vast, crashing, smashing, splintering, roaring and thundering avalanche of falling timber!'[15]

The slashed forest was left to dry until the weather was hot enough for the annual burn, the frightening climax of the pioneer's year. It was sometimes hard to get a 'good burn' because of the heavy rainfall and the inability of wind to penetrate the thick scrub.[16] Farmers therefore chose the hottest summer days for these burns, 'the windier and hotter the day the better for our purpose'. These settlers of the world's most fire-prone forests awaited the most fatal days. They were unwittingly re-creating the natural process that had produced and sustained the very forests they wanted to destroy. And so the forest fought back with unexpected vigour and, as

Settlers cleared the forest with fire.

historian Warwick Frost has shown, secondary clearance greatly increased the cost of farm establishment and, in perhaps a third of cases, led to the abandonment of properties.[17]

The annual burn was 'a spectacle that for awful grandeur beggars description'. Everything was tinged with 'that weird, eerie, livid, yellowish-green hue … the face of the sun appearing like a great dull copper disc'.[18] It was a time of great unease and irritability as well as excitement, as residents breathed in their neighbours' smoke and anxiously watched the wind direction.[19] Laboriously constructed split rail fences, homes and sometimes human lives were lost in the cause of clearing. After a good burn logs were picked up and fired a second time to clear the ground. 'What a change two hours of fire had wrought! We were forest dwellers no longer', exclaimed one selector. W M Elliot rejoiced that 'not a vestige remains of the vast forest that once so stubbornly resisted our labours. Hill and vale covered in verdure as far as the eye can see!'[20]

A 'good burn' could so easily become a wildfire and on Red Tuesday 1898 it did. Stephen Pyne called this fire 'not so much an unprecedented aberration as it was a macabre, reckless parody of frontier land clearing, transience and violence'.[21] After all, it was the clearing strategy of settlers to transform 'timber to tinder', and there was always a backlog of drying, slashed undergrowth and nicked and ringbarked trees. In 1898 settlers who had already had so much to do with fire, whose work each year climaxed in flame, encountered a phenomenon quite outside even their experience: '[T]he sky began to take on an aspect so dreadful and threatening that it made one almost afraid. Its colour was a strange shade of purple, tinged with blood.'

The next morning, recalled J Western, 'the forest had largely disappeared, and through the murky atmosphere one could see the homesteads of neighbours half a mile away that we had never been able to see before'. So the 1898 Gippsland fire, in spite of its destruction of homes and human life, was welcomed by many as it 'opened up the country a great deal'. It was the ultimate pioneering weapon run amok and it accelerated the pace of settlement.[22]

Many selectors were misled into believing that tall forests guaranteed rich and fertile soil. Such beliefs could influence the pattern of selection. Dense forests were among the factors that drew selectors to the Warragul district in the early 1880s, ensuring that Warragul developed faster than Drouin and Waterloo (later Yarragon), even though all were on the new Gippsland railway.[23] By removing the forest, selectors unleashed new problems. On the steep erosion-prone farms of the Strzeleckis, children mud-sledged to school. When botanist J W Audas passed through the Neerim district in the early twentieth century he observed that:

> ... for many years axe and fire have been ceaselessly at work, and the once
> densely-clad hills have on every side been mercilessly denuded of their
> timber. 'You *must* cut it down to get *grass*', said one farmer, in surprise
> at my expressions of regret at the passing of the forest. Grass, yes – and
> bracken, too ...[24]

Bracken and scrub reclaimed clearings, and insect plagues attacked newly sown paddocks. Wild pigs and horses, cockatoos, kangaroos and wallabies, possums and even wombats harried the selector.

In his book, *Ecological Imperialism*, historian Alfred Crosby reminded us that European immigrants did not arrive in the New World alone, but were accompanied by 'a grunting, lowing, neighing, crowing, chirping, snarling, buzzing, self-replicating and world-altering avalanche'.[25] Domesticated animals, pests, pathogens and weeds were an awesome accompaniment of the colonisers, and were sometimes consciously introduced for industry and ornament and sometimes were an incidental dimension of the human invasion. Animals and plants were introduced into new colonies to stimulate and diversify the economy, for recreation and hunting, and to cater for European nostalgia. 'Acclimatisation societies' were particularly active in nineteenth century Victoria and the forests of ash were the scene of some of their experiments.

The Acclimatisation Society of Victoria was founded in 1861 with the aim of introducing exotic plants and animals for dispersion to different parts of the colony and to *improve* local nature. Crop plants and grazing animals (such as the angora goat and the alpaca) were imported, and fish, birds and animals were let loose in the wild. As well as land at Royal Park in Melbourne, the society had a depot at Phillip Island and one in the foothill forests at Gembrook. This 259 hectare messmate and stringybark allotment in the hills was used as a game-breeding establishment in the 1870s and 1880s. Clearing was soon begun. A close slab fence enclosed the property, and a paddock was prepared for cultivation. Game birds and species of deer were released onto the property, and almost 300 pheasants were sent up to Gembrook each year. Large numbers of Californian quail were introduced and bred.

Branches of the Acclimatisation Society sprang up in country districts. There was even one in the new, isolated gold town of Woods Point in 1865. Later in the century, when enthusiasm for acclimatisation was waning, the Gembrook property reverted to the Crown, and £1000 were paid for 'improvements'. Later again it was divided up and sold, but a small section remains today as a flora and fauna reserve. The earlier reserve placed nature in the power of humans; today's reserve offers humans an experience of nature.[26] It is a parable of changing attitudes to wildlife in the forests.

Several exotic species were introduced to Victoria through the forests of ash. The Acclimatisation Society released brown trout in the Watts River and Olinda Creek in the early 1870s, and further distributions of trout fry were made to streams in the Gembrook, Beaconsfield, Narbethong, Taggerty, Mansfield, Yea and Mount Disappointment districts throughout the 1880s.[27] The axis deer of Japan was first liberated at Sugarloaf Hill, beyond Yan Yean, in 1862, and more were set free at Yering (also at Longerenong) in 1870. Fourteen of the Acclimatisation Society's fallow deer went to Bunyip in the 1860s, and the number at Phillip Island increased to 80. Eight hog deer were set free at Gembrook in 1870. Sambar deer, Australia's most successful deer transplant and now frequently hunted, were first liberated at King Parrot Creek in 1863 and near Tooradin in 1868. Deer spread quickly through the ranges and became a problem to farmers from as early as 1910.[28] The 1939 fire was probably responsible for the spread of Sambar deer into the ranges and throughout Gippsland. Fires and logging created new and suitable habitats for them.[29]

Rabbits arrived in the forest later than in many parts of Victoria. They spread northward and westward across the colony in the 1870s and arrived in numbers in the upper Goulburn Valley in the 1880s. There were elderly people living in Alexandra in the 1950s who remembered their surprise as children at first seeing a rabbit. At Neerim the 1898 fire was a boon to many settlers, who sowed cocksfoot grass and watched it grow luxuriously in the absence of rabbits.[30] Rabbits began to invade Gippsland in large numbers early in the new century, and they reached plague proportions there during the First World War when much farmland was left to run wild while soldiers served abroad. Soon the rabbit really was everywhere. 'He has', it was reported solemnly, 'been seen on the summit of Mount Hotham, in winter'.[31]

But the destruction wrought by rabbits in this region from the late nineteenth century has overshadowed the earlier era of settlement in the forests when indigenous animals and plants were the farmers' most hated assailants. As Warwick Frost has argued, indigenous pests were more of a problem than exotics on the 'wet frontier' of the great forests: 'Studies of European settlement in Australia have tended to focus on exotic pests', he argues, for these were the invading opportunists adapted to colonising disturbed environments. But in the wet sclerophyll forests, indigenous flora and fauna were also adapted to dramatic, even catastrophic disturbance and so 'it was eucalypts, wallabies and dingoes, not rabbits and Scotch thistles which caused the greater difficulties for the struggling selectors'.[32]

The annual burns and the seeding of paddocks for stock were followed by years of battling marauding wallabies, wombats, dingoes, eagles, cockatoos and plagues of insects. In the cleared forests of South Gippsland, 'the grasshopper ... regularly every Autumn devoured and laid bare the grass paddocks'. Selectors recalled 'the scourge of caterpillars':

I have seen a beautifully green paddock eaten out and left bare in 48 hours. And I have seen the caterpillars so thick against a big log that had stopped their march, that you could easily have taken a shovel and filled a barrow in a short time.[33]

Remnant bush and adjoining areas of state forest were suspected of harbouring this vermin, and so the settlers' hatred of trees was related to their fear of the unruly native life such vegetation protected. But once a cleared farming landscape was established, the way was clear for the exotic pests to invade the new, fragile grasslands.

'Improvements' – clearing, enclosure, residence, cultivation – demonstrated commitment to the land. One came to own it, legally as well as morally, by changing and subduing it. The *Land Act, 1869* was the most successful in settling farmers. Under the Act, men and single or widowed women could select up to 320 acres (130 hectares) of land under licence for a period of three years, and then become eligible to lease it and ultimately to purchase it freehold. Commitment to the land was measured by requirements to live on it, pay two shillings per acre rent, and make 'improvements' to the extent of £1 per acre. A portion of each holding had to be cultivated. The Lands Department stewards and bailiffs administered a system that put a value on ringbarking, grubbing, fencemaking and so on.

In the forest the rate of turnover in selection blocks was high, and many were neglected before they changed hands. This made the arithmetic of 'improvement'

'Afternoon tea in a Gilderoy Giant Tree', photographed by Nicholas Caire about 1903. Gilderoy is near Powelltown. (National Library of Australia)

especially complex. George Spicer, later the proprietor of the Mount Myrtalia Hotel at Gilderoy near Powelltown, sought to acquire a grazing lease near his farm in 1892. It had been forfeited by the previous lessee, Edwin Watters, whose 'improvements' had to be valued and were a subject of debate. '[T]he last lessee only scrubbed about 10 or 12 acres and fired it but did not pick it up', wrote Spicer to the Secretary of Lands. 'It has since grown and to a successor is not of much value. The trees rung on the same area were rung when the last lessee took possession.'[34]

The moral dimensions of farm settlement emerged clearly with the *Settlement on Lands Act, 1893*. Inspired by the Arcadian notions of the Reverend Horace Finn Tucker of South Yarra, it provided for the establishment of 'village communities, homestead associations and labour colonies'. It was a response to the poverty and un-employment of the depression years and was inspired by the belief that the country-side would restore the failed city dweller to innocence. People with little or no experience of the bush were sent into some of the most difficult areas for farming.

Many of the village settlements were established in West Gippsland and the Dandenongs. Of the three communities founded in the Dandenongs – at Monbulk, Ferny Creek and Woori Yallock Creek – two were utter failures. At Trafalgar, further east, families were expected to be self-sufficient on blocks as small as seven acres (2.8 hectares). At Wesburn, near Yarra Junction, 22 out of the 50 farmers abandoned the settlement or had their permits cancelled.[35]

Thomas Hughes of Adelaide took up land in the Neerim district under the Act. He was a single man and lived on the block. He immediately set about making the 'improvements' required by the Act. In 1895 he was clearing the land, ringbarking, grubbing, sowing grass seed, planting potatoes and fruit trees, and putting up fences. 'I am working on the land nearly all the time', he wrote to the Secretary of Village Settlements in 1896. 'And did grow plenty to support me could I have got it to any market I am in great distress for want of clothes in fact everythink but food'. He worked the block until his death in 1913.[36]

The government supported the village settlers with grants. In 1897 Hughes applied for £5 'to buy a horse some iron and seeds and some glass which I stand sorely in need of'. In the same year, John Berry, also of Neerim, sought £6 for a plough, 'Galvanised Iron for Roof and Chimney as the Wooden Fireplace is very dangerous in Summer', and some spouting to catch water. These were the humble instruments of 'improvement'.[37]

During the First World War the yeoman ideal found further legislative expression. Land settlement was adopted as the major repatriation scheme for returned soldiers, and in 1917 the Discharged Soldiers Settlement Act was passed. The failure of many earlier, closer settlement schemes was ignored in the prevailing spirit of gratitude to the soldiers. Suitable agricultural land was hard to find, and surveyors were instructed to squeeze as many farms as possible into the one estate. A real effort was made to return soldiers to their home territory. Since many came from Melbourne, there were strong concentrations of settlement in the dairying and market garden areas east and southeast of the city, in the foothills of the ranges. In general, the farms were too small and overvalued, and insufficient time was left for the settler to get established before repayments were due. The failure rate among soldier settlers was high, especially at the forest front where many lost the battle against bracken.[38]

The most inaccessible mountain forest was retained as Crown land, and some of it leased for grazing. Holders of grazing licences were only interim tenants pending the selection of land for cultivation. Sometimes they selected land themselves, as a way of securing portions of their large runs. Cattle could obtain scattered grazing in a burned forest. Forest officers approved grazing leases in thickly timbered mountain country, and the mature open forests were the best for this purpose. 'I beg to report that I don't know of any objection to this area being licensed for grazing purposes; it is old matured forest', wrote W D Ingle of applications for a grazing area in the Upper Yarra.[39] The Keppel family retained a grazing lease for 80 years until 1963 in the mountain forests north-east of Marysville. On the Baw Baw plateau, a subalpine environment with some open snow plains, cattle were first grazed in the late 1890s.[40]

Grazing in a mountain ash forest, however, was less common than grazing in the drier foothill forests below and the alpine ash forests above. There was much less grass there. The mountain ash belt is characterised by an understorey of ferns and shrubs, not conducive to grazing. Also, the mountain ash did not respond well to burning, although cattlemen claimed that it did not harm the saplings greatly: '[I]t would only amount to a natural thinning'.[41] Particularly in the drier forests, graziers used fire as the Aborigines had done: to keep the forest open, to clean up the scrub, to encourage a 'green pick', and to protect themselves and their stock from wildfire. In autumn a portion of each run was burnt. It was a tradition handed down over generations, sanctioned by long usage. It was this habit of burning that generated increasing government opposition to the cattlemen and women. They were blamed for many forest fires, for letting 'the red steer' run through the bush. A V Galbraith, Chairman of the Forests Commission, described the cattleman as 'the scourge of the forest'.[42]

The grazier and his firestick had long been criticised by forest managers. In 1914 the State Forests Department of Victoria again raised the matter in its annual report:

> Offenders do not choose a frequented place, or time of day when foresters
> or patrol men are abroad, to fire a forest, and if two men are together on
> cattle runs 'in good burning weather' they generally separate before using
> the match.[43]

Burning was an accepted practice, publicly defended, but individual acts were clandestine.

Not always, perhaps. A E Kelso, engineer in charge of water supply for the Board
of Works, recalled travelling with an employee of a grazier one hot February who kept
lighting matches and throwing them, while still burning, into the bush alongside the
track. Asked why he was doing this, the man replied that it was a signal to his boss
that they were on the way.[44]

In the early 1920s attempts were made to control or remove cattle grazing from
the forests. The newly formed Forests Commission of Victoria (founded in 1918)
assumed control of the State Forests and forced most graziers out when they were not
prepared to give an undertaking not to burn their leases. This policy strengthened
following the disastrous 1926 fire.

Cattlemen, however, believed that their burning kept the forest safe from fire by
keeping fuel loads down. 'Every bushman knows that you must burn country to make
it safe', exclaimed John Cameron, a Mansfield grazier. Cattlemen continually pointed
to the two devastating decades of fire that followed their removal from the state
forests, and argued that it was the Forests Commission's lack of burning that had
allowed the 1926, 1932 and 1939 fires to happen. 'If I had a manager and he
had three bad crashes in 13 years, I think I would sack him', declared cattleman
Frank Lovick to the 1939 Royal Commission. Another grazier, John Findlay, had
the Blue Range run between the Rubicon and the Little River. He told the 1939
Royal Commission:

> There is a beautiful forest at the head of it. I called on Mr McKay [of the
> Forests Commission 1918–24] and said 'If you do not burn that forest, you
> will lose the lot.' … When all the stockmen were in the bush we burnt those
> forests and none of them were killed. Since the Forestry officers have taken
> charge of it they have had it for practically twenty years, we have had two
> bad fires and this one has burned from one end of Victoria to the other.

The Forests Commission, according to the cattlemen, allowed the forest floor
to get 'too dirty'.[45]

There were vested interests here, of course. The Forests Commission, some said,
was blaming the cattlemen in order to extend its power over the Crown lands they
occupied. The cattlemen were defending their livelihood. In front of the 1939 Royal
Commission they rarely admitted to the value of 'a green pick' after fire. After

hearings of the commission in Mansfield Judge Stretton claimed that '[m]uch of the evidence was coloured by self-interest. Much of it was quite false. Little of it was wholly truthful.'[46]

Judge Stretton was able to pursue the cattlemen in a subsequent enquiry. In 1946 the Victorian government appointed a Royal Commission to enquire into the grazing of forest lands in Victoria, particularly in relation to water catchments and timber-producing areas in mountainous regions. Furious dust storms in the 1930s had alerted Victorians to the alarming problem of soil erosion, and following another Royal Commission a Soil Conservation Board was established in 1940.

During the course of the 1946 enquiry, Stretton investigated the extent to which grazing accelerated soil erosion or reduced water catchment efficiency. Stretton's report again sheeted home the blame to the graziers of upland forests: 'It is considered by many', he reported, 'that grazing is more harmful to the forest than is timber-getting.' It was not just that they 'burned with an untroubled conscience'; their cattle also increased the amount of bare ground and soil erosion. In a memorable phrase, he indentified 'an inseparable trinity – forest, soil and water'. Stretton condemned the way forest management was divided between the Department of Lands and Survey and the Forests Commission and recommended that control of forest grazing be vested in the Forests Commission. The Department of Lands and Survey, he thundered, 'is not interested in land welfare. Its real interest is in land transactions.'[47]

Many farms and grazing leases were swallowed up by the forest. In the 1880s Henry Middleton found his 860 acre (348 hectare) grazing lease in Smyth's Creek watershed near Warburton too big and inaccessible to 'improve'. He managed to scrub 50 acres (20 hectares) or so. It was no more than 30 acres (12 hectares) according to the Lands Department, and not enough to warrant continuation of his lease. The forest also defeated his successors, John Kefford, who followed him in 1891, and Thomas Linsley in 1898. In 1909 the land was withheld for future reservation 'as it commands the best means of access to large areas of good Crown timber in the permanent forest reserve on Smyth's Creek watershed.'[48]

In the La Trobe and Loch River valleys near Noojee, land carrying mountain ash, messmate and silvertop with a particularly light scrub growth was opened for selection in the 1880s. The land was cleared, burnt and cultivated, but within ten years nearly all the selections were abandoned. The West Australian gold rushes had

LYREBIRD

RORY O'BRIEN

The Superb Lyrebird *Menura novaehollandiae* is unique among Australia's birds and is endemic to the cool, temperate forests of south and central eastern Australia. It was introduced from mainland Australia to Tasmania in 1934. The Superb Lyrebird is one of two lyrebirds in Australia (Family: Menuridae); the other being Albert's Lyrebird *Menura alberti*, which is restricted to south-east Queensland.

When Superb Lyrebirds were first encountered by white settlers in 1797 near Sydney they were thought to be pheasants. It was not until 1848 that John Gould considered the lyrebird to be related to the songbirds (Passerines). That the lyrebirds are songbirds is quite evident by their syrinx (the vocal organ), but they differ from other songbirds in having three (rather than four) syringeal muscles, which are used to control the output of song. Furthermore, on the basis of genetic studies, the lyrebirds are foremost related to the scrub birds (Atrichornithidae), also found in Australia, and overall to the crows and honeyeaters, other songbird families.

The most distinctive feature of the lyrebird is its tail. The tail is longer in the male and shorter and duller in the female. The tail comprises 16 feathers in total, of three types: filamentary plumes, lyrates (a pair) and wires (a pair). The male uses its tail when displaying to females in the breeding season by 'dancing' on an exposed mound on the forest floor. In addition, the male mimics other forest birds to attract females, often using his song in combination with tail displays. A single female is usually attracted to the male's song and display, and males can mate with several females during the breeding season (polygamous breeding). The female alone builds a large stick nest, usually low down in the canopy, and lines the nest with fibrous roots. A single dark brown egg is laid. The peak of the breeding season is in winter, but breeding can occur throughout the year.

The Superb Lyrebird performs among dense forest undergrowth and is therefore difficult to observe. For this reason its behaviour has seemed 'secretive' and this, combined with its beauty and mimicry, has promoted mystique, myth and also awe. Of the people who have sought to establish some of the facts of the lyrebird's biology, one of the most unusual was Tom Tregellas (1864–1938), who lived for several years in a hollow log, which he named *'Menura'*, in the forest at Kallista. Tregellas studied and photographed lyrebirds in the 1920s and 1930s and many of his glass negatives and lantern slides are in the collections at Museum Victoria. People have

Tom Tregellas was an accomplished photographer. He captured the habits of the Superb Lyrebird well, and with no interference to the birds. (State Library of Victoria)

also celebrated the lyrebird as a native 'icon' and have designed costumes, music and art embodying the bird's image (see pp. 126–7).

The Superb Lyrebird feeds on invertebrates in ground leaf-litter and prefers areas that provide easy access to food. In mountain ash country, its habitat is tall shrubs with bare ground beneath. Therefore, the lyrebird is sensitive to clearing or disturbance of the forest canopy. However, the largest threat to the survival of Superb Lyrebirds is predation by the Fox *Vulpes vulpes* and by cats and dogs. The introduction of lyrebirds to Tasmania, where the fox is absent, is an added insurance to the long-term survival of the species.

lured many settlers away from their unrewarding fight with the forest. The grasslands they hoped to establish were swamped by a much denser growth of scrub. Soon a vigorous young stand of mountain ash asserted itself over the scrub. By 1918 a later owner, Dr Samuel Peacock, had two sawmills operating on the land to produce case timber from the young forest. In 1926 a fire destroyed both forest and mills. In 1933, 2000 acres (810 hectares) of the regenerating ash forest were purchased by Herbert Brookes and George Nicholas and handed over to the Forests Commission to provide forestry work for hundreds of unemployed youths who, it was hoped, would gain in discipline and character through their outdoor forest work. 'Improving' was again at the heart of this rural project, but this time the moral ideals concerned itinerant bush work rather than settlement and were in the service of forestry rather than farms.[49]

As railways stretched out from Melbourne towards forest areas in the late nineteenth century, and a large market for timber became accessible, trees had their uses. The early development of systematic sawmilling can be discerned in the increasing respect for the timber that would-be farmers cleared from their properties. The timber industry helped support many selectors, and vice versa. Stewards of the Village Settlement schemes of the 1890s commented on the timber resources of the lands they parcelled out. Lands Department bailiffs noted warily how aspiring applicants for land 'follow the sawmills'.[50]

Some farmers who had spent their lives clearing began to doubt that they had also been 'improving'. The regretful tone of their reminiscences had its origins in nostalgia and aesthetics, a sense that in chopping down a forest they had lost something of wonder and grandeur.[51] However, the regret was also economic. M Hansen, a selector in the Great Forest of South Gippsland, reflected in 1920 that it 'was perhaps a grave error to destroy all this valuable timber. I am inclined to think that within a very brief period those who have saved a few acres of timber will find that it will be the most valuable crop the land has ever yielded.'[52] An old Dandenong farmer put it this way in the 1950s:

> On my holding, when I'd cleared it from big timber I thought I'd finished. But no fear! The little saplings came popping up, crowding out my turnips. I used to knock 'em over like walking-sticks. Yet, come to think of it, if I'd let 'em grow where they wanted to, I might have done better. Reared a stand of mountain ash that'd 've paid me better than all the turnips I've raised in forty years.[53]

4

CROSSING THE BLACKS' SPUR

On the edges of the tall forests, Aboriginal people also played the 'improvement' game but came up against a government concerned more to improve them. At Coranderrk in the Yarra Valley and Jackson's Track, north of Drouin and Warragul, Aboriginal people established communities with tenacity and goodwill and both were ruthlessly dismantled by white society, partly out of malice, greed and suspicion, and partly out of misunderstanding and misguided benevolence.

Coranderrk was a station administered by the Board for the Protection of the Aborigines and was established at Aboriginal insistence in 1863 and officially closed in 1924. Jackson's Track was an informal community of Aboriginal people that grew up under the protection of the forest and sympathetic white landholders from the 1940s to the early 1960s, when it was demolished by Drouin Council. Both settlements were, for a time, economically and socially successful, distinctively Aboriginal and also accommodating of European ways. Their destruction by the society around them illustrates the discomfort and suspicion with which governing Australians have continually confronted autonomous and productive Aboriginal communities.

The Black Spur as it is now known – a name that suggests the darkness and luxuriance of the vegetation, or the after-effects of massive fire – is one of the most celebrated sites in the forests of ash, a 'beauty spot' on the road between Healesville and Marysville where tourists can be completely enclosed by tall trees and ferns. When the forests of the Black Spur were revealed to be under threat from logging in 1889–90, there was 'an immediate explosion' in the Victorian parliament and many people wrote and spoke in defence of the grandeur of this particular forest, with its 'magnificent timber, its sassafras and myrtle gullies ... [a] fairy spectacle which we have too much neglected', 'one of the finest spots on the face of God's earth'. Edwin J Hart wrote in the *Argus* of how 'the barbarian' would have seen 'such woods as clothe the slopes and long ravines of the Black Spur ... [as] a living temple', and he questioned a civilisation that would, instead, seek to clear such forests.[1]

It was probably forgotten then, as Victorians still forget today, that the Blacks' Spur actually memorialises those very people who were perceived as 'barbarians', the Aborigines. The spur was named after the Woi wurrung and the Daung wurrung people who used it to travel between their territories and who, in the early 1860s, made the journey several times through the tall forests in quest of a permanent home, a safe refuge amidst the maelstrom of European invasion. The Blacks' Spur deserves to be remembered, above all, as a path of pilgrimage to a 'promised' land.

The story of the crossing of the Blacks' Spur is a history in microcosm of the European dispossession of Australia's indigenous peoples, but also a story with peculiar and iconic status. The people of the Kulin clans were continually shunted around by government officers and rival white landholders, pushed back into the edges of the forest, promised land and betrayed, and convicted even for their successes. They continually made direct approaches to the government or the press, articulating their needs and hopes; they tried to work the new systems of law and obligation that were being imposed upon them.

South-west of the tall forests, on the site of Melbourne in June 1835, John Batman met *ngurungaeta* (clan leaders) of the local Kulin clans and claimed he had negotiated a treaty with them, purchasing two parcels of prime grazing land totalling 600 000 acres with 20 pairs of blankets, 30 axes, 100 knives, 30 mirrors, 200 handkerchiefs, 100 lbs of flour and 6 shirts as payment for the first portion, and 20 pairs of blankets, 30 knives, 12 axes, 10 mirrors, 12 pairs of scissors, 50 handkerchiefs, 12 red shirts, 4 flannel jackets, 4 suits of clothing, and 50 lbs of flour for the second. More of these goods were promised as 'yearly rent or tribute'.

Such a treaty questioned the legal foundation of British settlement, but both the New South Wales and imperial governments refused to recognise the transaction on the basis that only the Crown had the right to extinguish native title. But the Kulin participated in what they probably saw as a version of their Tanderrum ceremony which traditionally acknowledged a sharing of rights in, and use of, territory, and

Aboriginal people later saw the treaty as a recognition of their land ownership. Somewhere in the background of this ambiguous ritual, this legal embarrassment, this confidence trick, was an 11-year-old Woi wurrung boy named William Barak who would spend a lifetime paying his respects, often uselessly, to European decorum.[2]

In the 1850s the intensification of land use in Victoria, following the discovery of gold and the rapid escalation of an immigrant population, prompted the colonial government to actively rationalise and constrain Aboriginal life. By 1863 the population of Kulin people had dropped dramatically from thousands to under 200. The Board for the Protection of the Aborigines was established in 1860 to intervene systematically in indigenous community life, forcing people to move onto a small number of tightly controlled reserves, only later to use its powers to expel some of them on the basis of colour and descent, as well as their degree of political resistance.[3]

From the 1850s the *ngurungaeta* of the Kulin Confederacy encouraged survivors of the European invasion to make a new home for themselves on an approved settlement in the region. Independent of the government protectors' activity elsewhere in Victoria, they formed indigenous alliances across the various Kulin peoples and petitioned collectively for some of their land to farm as compensation for the loss of their traditional resources.[4] The early missionary, James Dredge, had recorded in 1839 how Aboriginal people of the Goulburn district 'speak of their country to a stranger with emotions of pride'.

In 1859 two Woi wurrung headmen (Simon Wonga and Munnarin) and five Daung wurrung clan leaders (Beaning, Murrin Murrin, Parugean, Baruppin and Koogurrin) visited the official guardian, William Thomas, and requested a block of land to be set apart for their people close to a culturally significant site on the Acheron River near Buxton.[5] They assured Thomas that 'they would cultivate and set down on the land like white men'. Soon they were, in Thomas' words, 'wending their way to their Goshen (Promised Land)'.[6] Although they were granted almost 5000 acres (2000 hectares) and worked it enthusiastically, local white settlers, a number of whom were chosen to oversee the station as trustees, swiftly manoeuvred the 80 Aborigines off the desirable land and gained it for themselves.

John Green, a Presbyterian lay preacher who had ministered to the goldminers and the Woi wurrung along the Yarra since his arrival in the colony from Scotland in 1858, became a crucial recruit to the Aboriginal cause. In 1861 the new Board recognised his work by appointing him guardian at the Upper Yarra Camp of the Woi wurrung at Woori Yallock, and authorised him to distribute supplies to them. The following year the Board gazetted a new Aboriginal Station at the site on Woori Yallock Creek, but very soon there was pressure to move them from there also as miners' tracks to the new goldfields crossed and surrounded it. Another reserve on the Watts River, gazetted to replace the Woori Yallock site, was also being contested by an Upper Yarra pastoralist. Meanwhile, the Aboriginal community on the Acheron, having lost their land to the squatters all over again, were being shunted further into the mountains to a station called Mohican.

Green had secured Board permission to cut a new track through the bush over the Great Dividing Range into the Goulburn River (Daung wurrung) country to 'benefit the miners' who were now feverishly surging into the mountains. In 1862, Green's family, together with a party of Woi wurrung, blazed a track over the Blacks' Spur, probably following a traditional Aboriginal pathway between Daung wurrung and Woi wurrung territories, and attempted to settle at the official but unpopular Mohican station.[7] But the Kulin people had little heart for this further attempt at clearing, fencing and building homes, especially in what they considered to be 'cold country' and a site not of their own choice. They strenuously resisted this further dispossession and 51 of the 86 residents of the original Acheron Station refused to make the transfer to the Mohican. Yet by 1863, after four years of struggle, they were forced to abandon their preferred home in Daung wurrung country, land that 'they had settled on and selected' and that Thomas 'had promised them ever should be theirs'.[8] The Kulin people had lost both their Acheron reserve and their faith in European promises, 'all because [as Thomas put it] a needy broken down squatter owed for £800 to two merchants'.[9]

In February and March 1863 the Kulin people again crossed the Blacks' Spur in search of a 'promised land'. Their unusual party was noticed by miners and memorialised in the name of the range. It was a journey driven by both desperation and hope. It was a pilgrimage inscribed with Christian meanings, and one that was sometimes remembered, in both cultures, as a story of Exodus. Two leaders of the Wurundjeri people, Simon Wonga and William Barak, led over 40 Aborigines (many of them refugees from the Acheron scheme) through the cool temperate rainforest and giant ash to a site they had chosen as their home on the river flats between Watts River and Badger Creek, just south of present-day Healesville. Their passage

'The Yarra Tribe starting for the Acheron, 1862' This photograph by Charles Walter
commemorates one of the crossings of the Blacks' Spur by the Kulin people in search of a
'promised land'. (Museum Victoria)

through the bush, with children, cattle and an overburdened dray, and in the company of John Green and his young family, was slow but sure.[10] As Green reported to
the Board,

> [S]o strong was the desire for improvement by this time among the young
> men belonging to the Goulburn, that they all at once consented to leave, and
> go to the Yarra. After the young men consented, all the old ones consented
> also. And in the month of Feb'y when I started to proceed to the Yarra all the
> young men and two old ones started with me, and others sent their children
> as a token that they would soon follow.[11]

They named their chosen home, a reserve of 2300 acres (930 hectares), Coranderrk, a Woi wurrung word for the white flowering Christmas bush that grew in abundance there. Wonga and Barak had frequented this district since their childhood.

The Kulin people felt that this site, too, had been promised to them forever, even with the direct endorsement of the Queen through her representative, the Governor of Victoria, Sir Henry Barkly. In May 1863, the Governor held a public levee in Melbourne for all gentlemen to celebrate the recent marriage of the Prince of Wales as well as the Queen's Birthday. The leaders of the Coranderrk settlement decided to attend, equipped with gifts for Prince Albert. Simon Wonga, accompanied by Barak and a further 18 representatives from various clans, presented a written address, a speech in Woi wurrung, and presents of weapons and basketwork. They also spoke good English and had a few words with their host, the Governor, about their need for land. The Queen's letter of thanks arrived about the same time as the Coranderrk reserve was officially gazetted, a coincidence that encouraged the Kulin people to believe in the efficacy of their delegation, as well as the security of their home.[12]

By 1866 most of the surviving Kulin people had gathered at Coranderrk. The Kulin clans were all together for the first time since the great meetings in Melbourne during the 1840s. Simon Wonga made a speech to the Daung wurrung early in 1865: 'Mr Green and all the Yarra blacks and me went through the mountains. We had not bread for four or five days. We did this to let you know about the good word. Now you have come to the Yarra. I am glad.'[13]

The Kulin people regarded this farm and township as their own, and expected to manage it themselves, seeing Green and other officials as helpers rather than as guardians. John Green was almost unique as a manager of Aboriginal people in Victoria for his treatment of the Kulin people as 'free and independent men and women'. The historian of Coranderrk, Diane Barwick, described how Green 'never would treat the Kulin as anything but free men and women equal to himself in all but knowledge of European ways. He eventually lost his job because he upheld this principle'. 'My method of managing the blacks', he explained, 'is to allow them to rule themselves as much as possible'.[14]

The Kulin people built huts, cleared and worked the land, and produced and sold cloaks, baskets, boomerangs and other artefacts. Building, fencing, clearing and cultivation proceeded during the 1860s with a speed one official considered 'truly

'Hop Gardens at Coranderrk' photographed by Frederich Kruger in the 1870s. Hops grown at Coranderrk generally commanded the highest prices at the Melbourne markets. (Museum Victoria)

astonishing'. By 1867 they were grinding flour from their own grain and cutting timber at their sawmill. These Aboriginal farmers won gold medals at Royal Agricultural Shows and scarcely a year went by when hops grown at Coranderrk did not command the highest price at the Melbourne markets.

Traditional affiliations and ceremonies persisted. Coranderrk became a tourist attraction of the Upper Yarra, a showplace to take distinguished foreign visitors, and it also drew up to 2000 local visitors a year. The settlement was testimony to the resilience and adaptability of Aboriginal culture.[15] Survivors had certainly abandoned old patterns of residence and subsistence in order to assemble on this site and embrace European-style farming, but indigenous concepts of political authority and responsibility for land undoubtedly still operated. Most Europeans tended to see only the superficial changes and hence were unaware of the strength of these cultural continuities. But many were both amazed – and annoyed – by the determination of the Coranderrk residents to manage their own affairs.[16]

From the late 1860s, European selectors increasingly eyed land along the Yarra River flats, and this greed increased as the Coranderrk residents made a success of hop cultivation in the 1870s. The Board siphoned off profits to general funds. The Board had supported Green and had taken some pride in Coranderrk's success, but as Europeans increasingly coveted the land under Kulin control, government officials were again prepared to make light of Aboriginal attachment to particular land. John Green resigned under considerable pressure from the Board, and by 1875 the government had decided to sell the reserve and move its inhabitants to a distant site on the Murray River. 'The Yarra [is] my father's country', objected Barak in February 1876. 'There's not mountains for me on the Murray.'

This decision initiated more than a decade of 'rebellion' by the Kulin people, a word used by people at the time and one that has been commemorated in Diane Barwick's superb history of the Aboriginal campaigns entitled *Rebellion at Coranderrk*. When Barwick began to study the station in the 1960s, she found that the Aboriginal memories were strong and saw that 'a study of Coranderrk had considerable relevance for modern "land rights" protests throughout Australia'.[17]

Coranderrk was just one of many sites of Aboriginal rebellion that have for so long gone unrecorded in the written history of Australia. Between 1874 and 1886, the Kulin people rebelled against loss of land and autonomy. They learned of the proposals for their removal by reading the *Age* and the *Argus*. They campaigned determinedly for recognition of their right to occupy and control the small acreage they had farmed for two decades. Their vigorous defence prompted a Royal Commission in 1877 and a Parliamentary Board of Inquiry in 1881. Both recommended that necessary repairs be carried out and the reserve be made fully self-supporting. Coranderrk was also proclaimed a 'preserved reservation', a status which could be reversed only by Parliament. As historian Inga Clendinnen put it, 'For the moment, the land was safe. But not the people.'[18]

Barak, that patient supplicant to the invader's conscience, led the residents in a campaign of delegations, strikes, protests and letters to the press expressing their concerns to the Victorian government and public. He pleaded: '[G]ive us this ground and let us manage here ourselves ... and no one over us ... we will show the country we can work it and make it pay and I know it will.'[19]

The government denied that the residents had inherited any rights to land, argued that Coranderrk was 'not their native ground', continually questioned the authenticity of the protest by suggesting that the residents were being manipulated by outsiders, and further declared that, in any case, they were not really Aborigines.[20] In 1886, 60 residents were ejected from Coranderrk on the eve of a long, harrowing depression. The *Aborigines Act* of that year required all people deemed by the government to be 'half-castes' and who were also under the age of 35 to leave the stations.

Coranderrk was the major theatre of Aboriginal culture in nineteenth-century Victoria. Its proximity to Melbourne ensured that it was much visited and photo-graphed – photos of Coranderrk constitute the largest visual archive of an Aboriginal community in the nineteenth century – and, as Jane Lydon has shown, the Kulin people manipulated the visual imagery to their own purposes, just as they did the forms of European ceremony and statecraft.[21]

The sustained rebellion of its residents further brought it to the attention of the public. But the number of recorded visitors to the station – 'its nearness to popu-lation', its vulnerability to corrupting influences – was used by officials as evidence of the site's unsuitability in the long term. Also, the fact that a range of clans had gathered on this land was said to diminish their combined claim upon it: 'It was not at any time the head-quarters of a tribe'.[22]

Even the Kulin people's seasonal use of the tall mountain forests was used to undermine them in this battle over land. Pastoralist and Board member, Edward Curr, was dismissive of their attachment to the Yarra and its forests: 'They never lived in the mountains; they went there occasionally.'[23] When photographers placed Coranderrk people in studios with a backdrop of ferns, they were casting them against a regional natural icon, but also associating them with 'the fern craze', a primeval picturesque that was sweeping through middle-class drawing rooms in the late nineteenth century. And while Nicholas Caire was photographing the giant ash and celebrating them as 'the oldest inhabitants in the land' and lamenting them as 'a passing race', he was also photographing residents of Coranderrk. The forest was literally the backdrop to their lives, and their seasonal use of it was seized on as another symbol of their transience in this land. It was primeval and aboriginal, as they were. It was destined to be tamed and ordered, as they were.[24]

By the early 1890s, disease, dispersion and bureaucratic intervention and neglect had undermined Aboriginal autonomy at Coranderrk. Half the area of the reserve was excised for white farmers in 1893. This experiment in Aboriginal farming had not 'failed', observed historian Diane Barwick. It had ended because of policy decisions,

CORANDERRK CALENDAR

LINDY ALLEN

The Kulin people have a detailed and unique understanding of their traditional lands which include the tall timber forests east of Melbourne. To them this country enjoys an annual cycle of seven seasons and two non-annual seasons – the Fire Season that occurs about every seven years; and the Flooding Season that has a cycle of approximately 28 years.

To the Woi wurrung and Daung wurrung speakers, this is the 'cold country' where their clans' territories meet in the hinterland of the Great Dividing Range. This is the source of the Yarra and Goulburn Rivers, both major spiritual and physical markers of these clans. This is a landscape dominated by the main peaks of Tonne-be-wong, Toole-be-wong and Donne-be-wong, mountains renamed Mounts Riddell, Ben Cairn and Donna Buang respectively by the new settlers in the nineteenth century. Kulin people travelled through and around these mountains along their 'unmade tracks', as they were called by these settlers and miners.

Each of the seven seasons is recognisable by patterns of weather and movement of the stars coinciding with changes in the life cycles of plants and animals. The inter-relationship between all these elements form the seasonal markers used to determine when a particular event may occur or when it was necessary to relocate. During the coolest part of the year, the Kulin people lived in the mountains to shelter from the strong cold winds, and where an abundant supply of food was available from the hearts of *kombadik*, the tree ferns growing in the steep damp gullies.

---

1 *iuk*, eel season   (around March)

We used to set a line at night time, mostly in the lagoons. We got eels out of that … They'd go down to where Badger Creek runs into the Yarra … Eels were running that plentiful they'd make a net and put that in. **Martha Nevin, 1969**

This is a short season when eels are caught on their migration down the Yarra River. Hot winds cease and the temperature cools. Days and nights are of equal length. The star Canopus, home of Lo An Tuk, the ancestral hunter, is almost due north at sunset. Manna gums are in flower and Brushtail Possums are breeding.

2 *waring*, wombat season   (around April–August)

Wumangurruditj, that's wombat. Put a piece of wombat, a piece of pork and a piece of porcupine and you can't tell the difference. **Jessie Hunter, 1999**

Wombats become more active now and are seen basking in the sun, while male lyrebirds perform their courtship displays. Mornings are misty and the days cool and raining. The nights are longer. The hearts of the tree ferns provide the main source of food during this long, cold season.

3 *guling*, orchid season   (around September)

The orchids, that's those little ones. They grow down on the flat, too, across Badger Creek – has a little flower like a star. If we had a headache she'd pick that and boil it and give you a very little bit to drink. **Jessie Hunter, 1999**

The cold weather is coming to an end and it is the major flowering season. Orchids, flax lilies, murnong, native geraniums and the wattles are in full bloom. Pied currawongs sing and female tree goannas excavate mounds to place their eggs. Sagittarius rises in the southeast and Arcturus is seen on the northwest horizon soon after sunset.

*The straight, upright shoots of the Austral Mulberry (*Hedycarya augustifolia*) were used by Kulin people to make fire drills. This much prized plant thrives in the fern gullies of the wettest and most sheltered part of the mountain forests and was so valuable it was traded as far up the Murray as Lake Boga.* Coranderrk Aboriginal Station, c. 1900 (Estate of Professor F. Shaw, Museum Victoria)

4   *poorneet*, **tadpole season**   (around October)
You get a frog and tie his back legs and then put a great big cod hook … And you'd have to put a weight. Otherwise you'd see the frog back on the log … You'd catch a codfish that way. **Martha Nevin, 1969**
Tadpoles are prolific in the waterholes at this time as the temperatures rise and the rain continues. The days and nights are of equal length. Plants continue to flower. Frogs lay their eggs.

5   *buath gurru*, **grass flowering season**   (around November)
I reckon they must have done something with the seeds for damper to pulp down into a floury thing for damper. That's when the butterflies come to the flowers. They were food. **Brian Paterson, 1999**
The kangaroo grass and Christmas bush begin to flower. Orion arrives in the night sky and most of the orchid season plants are at the end of their flowering periods.

6   **Kangaroo apple season**   (around December)
Usually when a storm's coming, rain birds, black cockatoos, yellow crested, they come down from the mountains. And you can bet on it within two or three days it rains. Never fails. Two days and it rains. And they make a racket. **Brian Paterson, 1999**
The weather is warming and the Pleiades are rising in the night sky. Peron's tree frogs are laying their eggs. The kangaroo apple is ripe and kangaroo grasses, Christmas bushes and black wattles are in flower.

7   *biderap*, **dry season**   (around January–February)
We used to play around the tussock grass. Playing hidey. So they must have been big enough to hide around. And it was nothing to see a snake curled up inside the tussock grass. **Dot Peters, 1999**
When the valley is drying and eastern grey kangaroos are breeding, wombats are seen at night, native cherries are ripe and the growling grass frogs are restless. The Southern Cross is high in the south at sunrise.
In the fire season, a medium to hot burn from a deliberately lit bush fire or lightning strike burns part of the Yarra Valley. In the flooding season, the Yarra River rises over its banks and floods the Yarra Glen and Launching Place flats.

This seasonal calendar was produced by the Melbourne Museum in association with Woi wurrung and Daung wurrung people, the descendants of those Kulin people who relocated in 1863 to the campsite at Badger Creek that became known as the Coranderrk Aboriginal Station. Many of these people were pioneering families at Coranderrk and Healesville, and, a century and a half later, continue to live and work in and around the district and in Melbourne. While this cyclic calendar does not correlate directly with a 12-month calendar, it is shown here in relationship to the calendar year.

Distributing vegetables at Coranderrk in the early twentieth century. (Museum Victoria)

'and the stereotype of failure had originated in Board propaganda aimed at justifying the sale of Coranderrk and other reserves'.[25]

In 1924 Coranderrk was officially closed, and most of the remaining residents were shifted to Lake Tyers, far from the Yarra of their upbringing. In 1939, in the wake of the Black Friday bushfires, managers of the Healesville Wildlife Sanctuary, jealous of the adjoining Aboriginal land, complained that the remnant bush of Coranderrk represented a fire threat.[26] Yet, as the Royal Commission into the causes of those fires began to reveal, it was the removal and suppression of Aboriginal fire regimes that had itself increased the chance of dangerous wildfire.

In 1948 the Coranderrk Lands Act was passed to revoke the property's status as an Aboriginal Station, and two years later Coranderrk was handed over for the Soldier Settlement Scheme, and 370 acres incorporated into Healesville Sanctuary.

In September 1991, the Victorian government returned ownership of a small portion of that land, the Coranderrk cemetery, to the Wurundjeri people, a clan of the Woi wurrung. At this ceremony, a government minister, Tom Roper, received some 'improving' gifts: an axe, some knives and forks, some flour and a blanket. These were repayment for items given to the Wurundjeri people by John Batman when he bargained for their land in 1835.[27] In March 1998, the last freehold property remaining of the original station was purchased at auction by the Indigenous Land Corporation with funding from the Federal government.

On another edge of the mountain ash country, at Jackson's Track, an Aboriginal community established in the mid-twentieth century also discovered a brief period of autonomy and freedom only to be literally crushed by white officialdom. In 1936 two white men, Daryl Tonkin and his brother Harry, travelled east from Melbourne in search of good luck and a selection. They were also 'curious to see the great forests in Gippsland'. They ended up at Jackson's Track north of Drouin, 'where we knew the biggest trees in the world were still standing', and here they purchased some land and established a timber mill. In 1939 the great fire roared down upon them, turning the day into 'an eerie twilight', destroying their mill and reducing their shack to a pile of twisted and blackened corrugated iron on a bed of ash. 'The trees in the forest were black and leafless, but we knew all but the mountain ash were still alive and would soon thrive.'

As they worked the forest, the Tonkin brothers employed Stewart Hood, an Aboriginal man forced to live apart from his family at the Lake Tyers Mission Station. Word got around the Aboriginal communities that there was a sympathetic environment and plenty of work at Jackson's Track. And so, during the 1940s, many Aboriginal people made another journey over another 'Blacks' Spur' to another refuge on the edge of the tall forests and the Tonkins were joined by the Hood family and the Austins, Roses (Lionel Rose, the world champion bantamweight boxer, grew up here), Lovetts, Wilsons, Coopers, Coombeses, Nelsons, Bloomfields, Hayeses, Stewarts, Tregonings, Mobournes, Scotts, Montas, Bulls, Moffats and Mulletts from around Victoria and parts of New South Wales.[28]

'We were free at Jackson's Track', recalled Russell Mullett. 'Anybody could come and visit. Your relatives could come and stay and the Aboriginal Welfare Officer or the Housing Commission people or the Manager of Lake Tyers couldn't hassle us.'[29] People there were free to stay together as families, free to bring up their own children, free to worship and go to the local school, and free to work the forests and to get bush tucker as well as earn a living. Daryl Tonkin in his extraordinary memoir, *Jackson's Track*, reminisced:

Whenever a whitefella stranger came the place closed up behind a steel wall impossible to penetrate ... The only protection they had was the fact that

they lived deep in the forest on private property and the only hope they had
was that the Welfare might not see them and might forget about them.

Tonkin, who fell in love with an Aboriginal woman, Euphemia Mullett, was
impressed by the wisdom of his new, extended family in the face of such pressures:
'they seemed to have infinite patience, as if they had a deep knowledge of their own
strength and of human nature.'[30] But his love, his forest and even his private property
could not forever defend the Aboriginal people from the interfering strictures of his
own white society.

In 1957 the *Aborigines (Housing) Act* made houses in towns available for
Aboriginal people and they were encouraged and bullied to move into them by the
newly formed Aborigines Welfare Board, the purpose of which was to promote
the 'moral intellectual and physical welfare [of Aborigines] with a view to their
assimilation into the general community'.[31]

This new policy offered Drouin Council what Daryl Tonkin called the 'final
solution' to the 'eyesore' of the Jackson's Track shanties, which were now exposed to
visitors' eyes by a new road cutting through the camp. In 1962, residents at the Track
were given an ultimatum to move into a house in Drouin or be shifted back to Lake
Tyers, the only settlement remaining under government control and itself under
threat, and they were given a week to decide and pack up their belongings.[32]

On the appointed day, two trucks drove into the camp, one carrying a bulldozer.
People were loaded onto the back of them and, as they watched, their homes and
anything left in them were crushed and flattened, splashed with kerosene and burnt.
This large group of Aboriginal people was then dumped at a single, two-roomed
house in a paddock near the Drouin golf course, and provided with tents. They were
later separated from one another and dispersed to single houses in resentful, white
neighbourhoods.[33]

One hundred years after the Acheron Station was reclaimed by pastoralists calling
themselves 'trustees', 100 years after Coranderrk was founded and its hopes betrayed
and assimilation enforced, some things had not changed in the forests.

5

MINING

On Black Thursday, 6 February 1851, the skies of south-eastern Australia darkened and ash fell on ships at sea; a great bushfire swept across thinly populated Victoria as if to announce a new era.

George Gordon McCrae was 17 and working in a government survey party near Mount Macedon, west of the forests of ash. He awoke that morning to find the sky 'one dull leaden hue excepting near the horizon where it is red like port-wine'. The sun seemed 'like a ball of red-hot iron' and birds were dropping dead from branches; there was '[s]uch heat as I never experienced in my life before and never wish to again.' He and his companions waded into a nearby creek while the scrub along both banks took fire and 'our faces and bodies were next to roasting'.

There was no sleep that night, as great trees were crashing down all around them. Towards morning rain fell 'and the air became so chilly that I was glad of my blanket'. All across Victoria, 'from the western coast to the Australian Alps, from the Snowy River to the Murray', people fled and cowered before the fires. As William Howitt put it: 'The whole country, for a time, was a furious furnace; and, what was most singular, *the greatest part of the mischief was done in a single day.*'[1]

Within days of the great fire, Edward Hargraves publicised his discovery of specks of gold near Bathurst in New South Wales and released another social and environmental holocaust. Hargraves was not the only one looking for gold in early 1851. Did prospectors start the great 1851 fire, the first of such size since European

settlement? They certainly benefited from it. It opened up the forests, the rivers and
the outcrops of rock to their gleaming eyes.

Throughout Australia, before 1851, gold had not passed entirely unnoticed, for
shepherds had chipped specks out of rocks and turned the gold over in their hands.
Hargraves' strategic sense of publicity alerted colonists to the fact that gold could be
won systematically by washing the soil with pans and cradles. As valuable labour
flocked to New South Wales, Victorians felt the need to promote their own rushes,
and the suspicion of hidden riches and the offer of a government reward stimulated
local finds. One of the first Victorian discoveries of gold in 1851, made five months
after Black Thursday when Victoria was barely two weeks old as an independent
colony, was east of Melbourne on the Yarra River. Louis John Michel and his five
companions camped on Anderson's Creek near Warrandyte and found a rock with
veins of quartz. They panned the dirt in the creek and isolated enough yellow specks
to generate a rush of diggers. Despite this early eastern find, discoveries in the
Ballarat, Bendigo and Beechworth districts drew miners northward and westward
for most of the 1850s.

From late in that decade, however, gold was found in Victoria's eastern valleys and
mountains. The Emerald field was opened up in 1858, and there were brief-lived
rushes along the Little Yarra and Upper Yarra rivers. Hoddles Creek, Britannia Creek,
the Nicholson and Darling goldfields in the Yarra Valley and the Donovans and
Walshes Creek areas further upstream were among those places rushed in the late
1850s and early 1860s.

But the major goldfield near the forests of ash was further east, deep in the
mountains of the Great Divide. It became known as the Jordan goldfield, 'it being
a very hard road to travel'.[2] It stretched from Jamieson in the north through Woods
Point and Matlock to Walhalla in the south. Two events – Victoria's major 1860s gold
rush and the 1939 bushfire – bracketed this mining district's modern human history.
Settlements were founded on gold and annihilated by fire. Gold lured settlers beyond
the geographical barriers that had daunted their predecessors, and fire brought a
crisp end to communities already in decline.

The valleys of the vast Jordan goldfield were colonised from the north, temporarily
in 1854, again in 1857 and in a sustained way from late in 1859. By late 1862 parties
from Gippsland in the south had met those from the north in the quest for quartz
reefs along the Great Divide.

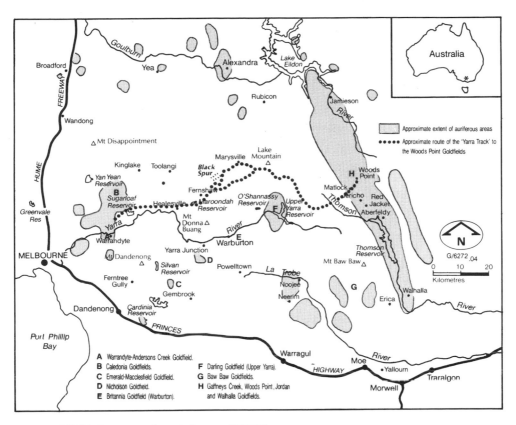

Major goldfields in or near the ash forests. (DNRE)

Jamieson developed from 1860 as a stopping place and market town for the new and rapidly extending goldfield. It was there that packhorses took over from drays and began their tortuous journey southward. At the height of the district's 1860s boom, over 500 horses and mules were carrying supplies and machinery to miners along the upper Goulburn Valley. The Woods Point field was opened up from the north in May 1861. During the next three years, spectacular quartz reefs (particularly the Morning Star) were quietly developed in this area while the bulk of the miners were seeking fast alluvial wealth in the broader Jordan Valley further south.[3]

It was the alluvial field in the Jordan Valley that revived Victoria's gold fever of the early 1850s. In November 1861 William Quinn, Owen Little, Gilbert Jones and Peter Sandford crossed the Matlock hill on the Great Divide and descended over 700 metres into the broad flat valley of the Jordan River. Here there was room for hundreds of miners. Nuggets were plentiful and sluicing was richly rewarded. From paddocks 12 feet (3.6 metres) square, yields of 50 to 120 ounces were easily obtained.

In February 1862, three months after the first parties had reached the Jordan, the Mining Registrar for the division reported some 4000 diggers on the field, most of whom had rushed southward from the upper Goulburn Valley. In 1863 the thousands of miners were reportedly sharing the valley with 2280 sluice boxes and 138 water-wheels. Jericho was the main settlement in the Jordan Valley, where 'you could, at any hour, hear the click of the billiard-balls'. It earned its name because tree ferns by the river reminded diggers of 'an illustration on a bible page showing Jericho with palm trees'.[4]

Initially this new extension of the Jamieson field was just that – tethered to civilisation by a 19 kilometre precipitous track over the Great Divide and then by a further 64 kilometres of packhorse trail to Jamieson. However, new tracks were quickly cut through the forest to the west and the south. McEvoys, Campbells, Bowmans and Porters tracks were cut in the early 1860s to give Gippsland traders access to the burgeoning Jordan Valley mining settlements. On one day in June 1862, McEvoy himself observed 40 packhorses following his path. The new tracks enabled the discovery of the Donnellys Creek field in late 1862, and by February 1863 Edward Stringer was spreading word of his find at Walhalla. At the same time, Hughes', Williams' and Sullivan's tracks were cut through the Upper Yarra to Matlock and Woods Point.

A more direct route to Melbourne was urgently required. From 1862 miners, packers and storekeepers opened up what became known as the Yarra Track. It passed through Healesville or Chum Creek and Fernshaw, up the Blacks' Spur, through Narbethong, Marysville and the Cumberland Valley, then followed the Great Divide to Matlock at over 1200 metres, from where it then descended to Woods Point. The country it passed through was a 'perfect jungle'. One traveller compared it with the delightful scenery of Ceylon. He enjoyed 'the sweet smelling freshness of the ferns', some of which towered over 10 metres above them.

Miners also encountered a forest ravaged and renewed by the Black Thursday fire. The same traveller in 1864 was puzzled by 'the dead tree country in the ranges'. Because of the whitened ramparts of spars the mountains at the head of the Yarra 'had the appearance of columnar basaltic precipices, not unlike the Giants' Causeway in Ireland.' A day or two later, near Matlock, he walked through 'one vast forest of dead trees, altogether destitute of bark or leaves; while the trunks and branches were bleached white with the weather.' He and his companions felt as if they were 'in a fog or mist'.[5]

The Jordan was soon eclipsed by quartz reef finds in the hills around Woods Point. The stunning yields from the Morning Star Hill reef – an average of over 10 ounces per ton in the March quarter of 1865 at a time when Bendigo mines were boasting an average of one ounce – prompted a mania of speculation that was to give Woods Point mines a bad name with Melbourne shareholders. 'In Bendigo', wrote one miner to the Melbourne *Age* in 1865, 'it is pennyweights to the ton but here it is ounces and pounds weight. Come one, come all.'[6] At Matlock, 13 hotels catered for the influx of diggers in 1865. Such was the nightlife that one traveller commented that 'on Matlock they turn night into day as usual and where the money comes from God only knows'.[7] New arrivals frequently commented on the presence of pianos and billiard tables in the mountain gold towns. They were registering surprise not just at recreational sophistication (the diggers played them well) but also at the sheer feat of getting them there.[8]

The forest did not yield its secrets easily. In 1863 Jack Russell stumbled upon a reef of glittering quartz that stood out of the ground like a wall. The story goes that he hung his billycan on a tree to signal possession and then returned to the mining settlement of Red Jacket. He never found the Billycan Reef again. Two years later, William Singleton and Fred Stander discovered a reef that they called the Royal Standard because of its bright colours. It rapidly became known as the richest mine in the Woods Point area, producing a sensational yield of over 9 ounces to the ton. It was undoubtedly the Billycan Reef, but Russell still believed that his find remained hidden and searched for it fruitlessly for 30 years.[9]

Quartz mining demanded considerable investment and technology to plunder the underground reefs, crush the rock and then extract the gold, often chemically. The isolated mountainous terrain and plentiful water shaped the distinctive features of alpine mining. Reefs were worked by means of shafts, adits (tunnels) or open cuts, and the quartz extracted was generally conveyed by chutes or tramways down to a battery or crushing site by a creek. Waterwheels drove the stampers, although crushing had to stop at the height of summer when water became scarce. Steam power was soon introduced, and trees were cut in their thousands to supply fuel. Tramways brought wood for the boilers, from increasing distances as the local forests were cut out. Timber was also needed to line the shafts and tunnels and to construct the battery frames and waterwheels. Wooden waterwheels proliferated as a means of power in the early years of the district, partly because they could be constructed from local timber and only the iron hub and buckets needed to be brought in.

Waterwheels are represented in the district's physical heritage. One has survived at Donnellys Creek, another is on display at the Upper Yarra Dam, and the remains of others indicate major battery sites in steep mountain gullies. Along Sailor Bills Creek (south of Jamieson), where mining continued intensively until as recently as the 1930s (and has now revived), a waterwheel trench survives as one of the few visible landscape features of the era. Water races also remain as prominent and intriguing relics. They were primitive but exact features of engineering, extending for many kilometres through rugged country and carefully maintaining a steady gradient. Water was needed not only to drive the wheels but to irrigate the batteries and sluice earth and tailings. Sometimes the races ensured clean water for domestic consumption in townships (such as Gaffneys Creek) where industrial sites hugged the creek.

Packhorses made their way tortuously along the tracks from Jamieson, Melbourne and Gippsland, carrying mining machinery for the new goldfields. Pianos and billiard tables were nothing compared with a boiler or battery. Drysdale and Company's battery in the Woods Point area cost £1300 for packing alone.

The arrival of mine machinery in an isolated mining settlement was a momentous occasion, as it represented faith and heralded prosperity. Cylinders, boilers and flywheels were sometimes carefully designed for the mountain journey and were cast in several pieces. One early observer recorded meeting a train of '50 horses each with a boiler plate across his back so enveloping him that only his head and tail could be seen'.[10] It was hard to get heavy machinery in, and just as hard to get it out. Some of it therefore remains at the bottom of mountain valleys, representing some of Victoria's most significant and intact engineering heritage. Boilers, flywheels, stamp batteries and steam engines have found a permanent home at the end of the Yarra Track.

Alluvial mining also left its mark. By the 1870s most rivers in the mountain forests had been prospected. Sometimes the legacy was dramatic: at Pound Bend, Warrandyte, and the Little and Big Peninsulas beyond Warburton, on the Jordan River at Jericho and on the Thomson River, massive tunnels were cut through rock and the watercourses themselves diverted. Miners sluiced the exposed river bed.

Shafts, adits, sluicing pits and water races deranged the rivers and eroded the river flats. Even the Yarra became muddier than its natural hue. In 1876, Victoria's Chief Inspector of Mines, Robert Brough Smyth visited the gold-fields beyond Warburton. He wrote:

> I visited all the mining claims on Starvation Creek. They are worked by hydraulic sluices, and great skill is shown in conducting these operations. The miners have cut good races, some of them many miles in length, and they appear to obtain, generally, very excellent returns for their outlay ... Pieces as large as a walnut are found occasionally.

A boiler being hauled to a mine near Gaffney's Creek about 1901. (From a print held by Mr F Cadan, reproduced from Brian Lloyd and Howard Combes, Gold at Gaffney's Creek, *Shoestring Bookshop, Wangaratta, 1981.)*

He concluded that this part of Victoria would be attractive to 'men who prefer moderate gains and independence, to a dreary and often unprofitable life in the towns.'[11]

Perhaps they were men like Sam Knott, who 'allus had wan at eleven'. He enjoyed a bit of sluicing at the bar as well as in the creek. Knott was a goldminer and an Upper Yarra man. Today he is claimed by every pub along the Warburton road and beyond. The full verse on the Carlton Brewery advertising poster that made him famous was:

I allus have wan at eleven
It's a habit wot's gotta be done
Cos if I don't have wan at eleven
I allus have eleven at one.[12]

'I allus has wan at eleven'. Sam Knott was an Upper Yarra goldminer who is now claimed by every pub along the Warburton road.

From the late 1860s the mountain hotels began to close. In 1872 the English novelist, Anthony Trollope, described the declining goldfield of Edwards Reef, north of Walhalla, as a 'miserable melancholy place, surrounded by interminable forest in which unhappy diggers had sunk holes here and there, so that one wondered that the children did not all perish by falling into them'.[13] At Woods Point the boom collapsed at the end of 1867, as the richer portions of the original reefs were exhausted. Little capital had been saved for further exploration. The goldfield drifted into decline in the 1870s. The working population came to depend on the few continuously operating mines: the A1, Morning Star, All Nations, Loch Fyne, Golden Lily, and Toombon, as well as the Long Tunnel at Walhalla. The largest yields came from Cohen's Reef at Walhalla, which from 1863 to 1914 yielded 1.5 million ounces of gold and paid £2.4 million in dividends.

By the 1880s about 2000 people lived in Walhalla, the centre of a field that was comparable to Bendigo and Ballarat in gold production. However, in terms of amenities, the town remained a frontier settlement and even by 1896 had no domestic water supply, court house or railway.[14]

The 1890s and early 1900s witnessed a widespread mining revival that continued until the First World War and, in many areas, overlaid and changed original patterns of workings. Gippslanders 'talked of nothing but these mines', and to say that one was 'born at Walhalla' was to claim membership of a select mountain race.[15] Anthony Trollope visited Walhalla in February 1872 and duly noted the 'pianoforte in the hotel sitting-room' and 'the billiard-table, – at which unwashed, earth-soiled diggers were playing, and playing, too, very well.' He wrote, 'I could not have believed there had been so much traffic across the mountains and through the forests, had I not afterwards seen the things at Walhalla'.

There were still trees on the hillsides when Trollope visited, and the town was picturesque and much photographed. Later in the century, however, the nearby hills became completely denuded of timber and were 'scarred and seamed with tram tracks and timber shoots'.[16] In 1896, when Walhalla schoolteacher Henry Tisdall revisited his old haunts, especially the fern-covered glades he remembered, he 'found that they were of the past'. 'The truth is', he reported to the Victorian Field Naturalists' Club, 'that mining and botany are incompatible.' 'The mines require firewood and timber, and the ruthless sawyer and woodsplitter has no veneration or respect for the beauties of nature.'[17]

Woodcutting was often known as 'that other gold mine', and at times in Walhalla higher wages were paid to the timber cutters who fed the boilers than to the miners who went underground. The wood-cutting industry at Walhalla was dominated by northern Italians, who formed a distinct community at Poverty Point, 5 kilometres upstream. They built bark houses in the bush with stone fireplaces (some of which remain), and for mortar they used the remains of termite mounds.[18] An extensive system of tramways tortuously carved long, gentle inclines out of the contours of Walhalla, and by 1899 was supplying the mines with up to 34 000 tonnes of wood a year. Such was the demand for wood that the Thomson River was bridged at Poverty Point in order to exploit the forests on the slopes of Mount Baw Baw. Even before the steel bridge was built in June 1900, ten further kilometres of tramway had been pegged out on the opposite side of the river. Today the bridge and tramway form part of the Alpine Walking Track.[19]

Lou de Prada, 'a man who was born at Walhalla' and a Poverty Point woodcutter, recalled the social as well as economic role of wood. Woodchopping was the endless task for children at weekends and after school; it was needed for the home fires, for cooking and warmth. When the police station woodheap was getting low, police arrested a couple of the town drunks and sentenced them to a day's woodchopping. Woodcutting was entertainment as well as discipline. Lou de Prada once said to a Walhalla resident, 'You have a good stack of wood there on your wood heap', and he replied, 'Yes, it's the only recreation we have.'[20]

Woodcutting was 'that other gold mine'. This impressive wood stack sat beside the Harbinger Mine battery house near Jericho in 1894. Ore from the main adit (mining tunnel) was delivered to the battery for crushing via the chute at the rear of the building. (Historic Places, DNRE)

The first decade of the twentieth century was one of Walhalla's most successful mining periods, but it was also the eve of its sudden decline. The Walhalla railway, built after decades of lobbying, was completed in 1910, just in time to take people, their belongings and some of their houses away. The Long Tunnel Extended and the Long Tunnel had employed over 300 men during the previous decade, but they wound up in 1912 and 1913 respectively. The last gold from these mines left Walhalla in 1915.

Mining continued quietly along the Jordan goldfield in the 1920s, and in the 1930s depression it received some impetus from the rising price of gold and high national levels of unemployment. Only two gold mines continued to operate after the 1930s: the A1 at Gaffneys Creek and the Morning Star at Woods Point. Both figured among the top ten gold producers in Victoria until well after the Second World War.[21] Other forms of mining plumbed and scraped the soils of the ash forests. Tin, Victoria's second most valuable mineral, was found at the head of the Bunyip River in 1876 and has been mined sporadically since. Wolframite, an ore of tungsten, was mined at Wilks Creek near Marysville between 1901 and 1919. Copper ore was mined at Coopers Creek on the Thomson River from 1863 until 1881, producing a total of about 800 tonnes of copper, most of it in the mine's final years.[22] The 1939 bushfire drew a final black line across the lives of the already dwindling mining communities.

Fire had opened up the forests to miners, and in the end it chased most of them out. Some of the forests logged in the Woods Point district after the Second World War began with the mining fires of the 1860s, and all around was the 1939 regrowth.[23] Walhalla's scarred and denuded valley is now thickly vegetated, and the old tramway tracks are embowered. Lou de Prada found the changes both beautiful and depressing. 'The period of the intrusion of man is over', he wrote, 'and an overbearing silence prevails over the mountains'.[24]

GEMS IN THE FOREST

BILL BIRCH

One of the best-kept secrets of the mountain forests east of Melbourne has been the existence of precious gemstones in the streams draining through them. Some of the geological history of the region can be read in the mystery and beauty of the gems.

The forests clothing the eastern ranges soften a landscape that can be traced back tens of millions of years. The oldest rocks are mainly siltstones and mudstones, deposited layer by layer in an ocean basin in the Silurian and Devonian eras. After being uplifted and squeezed into folds, the sedimentary rocks were invaded by magma late in the Devonian era. Gigantic volcanic eruptions poured layers of ash and lava into a large irregular caldera. These rocks now form the Dandenong Ranges. Immediately afterwards, two great plumes of granite magma intruded the sedimentary rocks. Long ago unroofed by erosion, the granite masses lie exposed in the Lysterfield Hills and in the region east of Gembrook.

During the cooling of the eastern granite mass, known as the Tynong Granite, pockets of hot watery fluids had accumulated amidst the crystal mush. In the ultimately solid granite, these pockets became cavities lined by perfectly formed crystals of quartz and feldspar. Scattered among them were black tourmaline crystals and even some colourless to pale blue topaz crystals. In places, there were veins of black crystals of tin oxide, or cassiterite. These natural jewel boxes lay undisturbed for millions of years, but erosion eventually removed the overlying rocks and exposed the crystal cavities to the weather. Like sluice boxes, the local streams sorted the crystals, light quartz from dense cassiterite, large crystals from small.

About 20 million years ago, volcanoes began erupting in the region again. These were much smaller than those that formed the Dandenongs and of a different type, giving rise to basalt lava. Some of the earliest eruptions released billowing clouds of crystal-laden ash. Among the crystals were small sapphires and zircons, which fell to earth to be quickly carried into nearby streams by storm rains generated by the eruptions. Lava flowed down some of the stream valleys, burying their sand and gravel beds. In some of the flows, small gas cavities provided space for silica-rich solutions to accumulate. These ultimately solidified into varieties of chalcedony, such as carnelian and agate.

As soon as the eruptions ceased, the endless processes of erosion continued, working on the new lava flows, as well as the old granite and sedimentary rocks. The streams

Waterworn sapphire crystals (to 1 cm) from the Pakenham district. (Museum Victoria)

now had more gemstones to work with. Sapphires, zircons and agate from the young volcanic rocks were added to colourless, smoky brown and amethystine quartz, topaz and tourmaline from the old granite.

For a brief period in the 1850s and 1860s, the colony of Victoria was being extolled as potentially the world's richest gemstone region. However, as alluvial gold and tin mining declined, so did gemstone discoveries, and by 1880 the excitement had long since dissipated. But a century later, the new hobby of lapidary prompted gem fossickers to retrace the steps of the early miners, and to sift the stream beds once again.

6

TIMBER TRAMWAYS

Timber tramways, which first snaked through the forests of the mining towns, came to typify the bush sawmilling era in Victoria's mountains. From the 1880s to 1950, a particular system of timber-getting and milling developed in the forests of ash that involved the establishment of isolated and temporary sawmill settlements deep in the bush, linked to civilisation and the Victorian railway system by long, narrow tramways. For the first half of the twentieth century, these forests were the focus of Victoria's sawmilling industry. The tall forests were inhabited; they were places of work; and the fragile lifelines of the bush communities were the timber tramways.

A network of sawmilling tramways did not become established in the mountain forests until late in the nineteenth century. At first, the bulk of Victoria's sawn timber was supplied by imports, some from Launceston and Sydney, most from the western coast of North America. Goldmining created an intense demand for local supplies of timber, and the forests near the mining towns were the first to be systematically exploited.

Sawmills were established in the 1850s in the ranges north and east of Melbourne to cater for the city's rapid increase of population and demand for building materials. John Wood Beilby built possibly the earliest sawmill in the central mountain forests at Fern Tree Gully in the early 1850s, and he found that timber shipped from Tasmania could be sold far more cheaply in Melbourne than timber brought by dray from the Dandenongs. However, Emerald and Belgrave became early centres for

Horses hauling logs on a tramway in the ash forests, 1918. (Historic Places, DNRE)

timber-getting in these eastern ranges, and sawmills were also established on the southern slopes of the Kinglake forests.[1]

The earliest exploiters of the ash forests were the splitters. Ash timbers were eagerly sought for palings and shingles because they split so easily, and many of the forest giants fell to paling splitters. A lone splitter could wreak considerable havoc. Even by 1874 the splitters had wrought 'immense damage' in small areas of the Victoria Forest in the Upper Yarra Valley. In that year it was estimated that for each tree used by a splitter, seven others were felled and abandoned. These men were said to have 'epicurean tastes'.

By 1898, nothing had changed. A splitter in the Toolangi district testified that he would often reject 70 per cent of the trees he felled.[2] For most of the nineteenth century, timber-getters used the forest under cheap, quarterly, fixed rate licences and did not have to pay royalties on the timber they extracted. The licence system produced a scramble for the best trees in a forest, leading to splitters and sawmillers 'shepherding' areas by felling and abandoning trees and frustrating competitors by obstructing tracks with fallen timber.

Relics of their raids endure in forests that survive. They tested trunks by chipping to see if a tree was fissile enough, and surface fires sometimes etched and preserved these signatures in charcoal. In some forests that escaped later sawmilling or massive fire, there are still the trees that the splitters felled and discarded, and the brown rot that rendered some logs useless for palings also left a lignin (a residual binding compound in wood) that preserved them.[3]

In Victoria a small group of enlightened bureaucrats and scientists – Clement Hodgkinson, Charles Whybrow Ligar, Robert Brough Smyth and Ferdinand von Mueller among them – argued the case for forest reserves from the 1860s.[4] But, as Tim Bonyhady has observed, '[W]hen Australia's colonial governments began to enact laws to limit the cutting of native forests, they always sought to protect new growth by setting minimum size limits and ignored the old'.[5] The dominant concern of governments and foresters was to maximise timber production and hence revenue to the Crown. The first Victorian legislation to create reserves for 'the growth and preservation of timber' was embodied in the *Land Act 1862*. This provision was strengthened in the 1865 Act, and within ten years, over 400 000 hectares of Victoria had been set aside as state forests or timber reserves. In 1870 special regulations were made to protect the Dandenong State Forest. These prohibited the unlicensed removal of timber, indigenous shrubs, fern trees or fronds of fern trees, and the stripping of bark, and limited the lighting of fires.

A series of reports from the 1870s onwards, detailed huge waste in the timber industry and irresponsible and ineffective management of the forest resource.[6] The main centre of scientific forestry in the British Empire was India, and European-trained Indian forest officers were harsh commentators on Victorian government policy in the 1880s and 1890s. Forestry Bills were presented to Parliament in 1879, 1881, 1887 and 1892, but lapsed. Forestry was the 'Cinderella of the departments', being placed at different periods under the control of the departments of Lands, Mines and Agriculture, just to underline the political priority of settlement.[7] The first Victorian Conservator of Forests, George Perrin, appointed in 1888, worked within a bureaucracy that opposed him.

These circumstances eventually culminated in a Royal Commission on Forests, which sat from 1897 to 1901, produced 14 reports and led to the *Forests Act 1907*, which established a Department of Forests. This professional emergence of scientific forestry consolidated with the establishment of the School of Forestry at Creswick in

1910. In 1918 a further Forests Act created a Forests Commission and established a fixed Forests Fund, which ensured that a substantial amount of forest revenue would be used for the improvement of the forests. This was in response to situations such as that in 1912 when the Victorian government obtained £55 000 in revenue from the forests but spent less than £3000 in the same year on reafforestation and forestry staff.[8]

The development of forest management coincided with the early years of intensive commercial sawmilling in the eastern forests near Melbourne. Much of the momentum for this legislative change came from the devastation and exhaustion of forests in other parts of Victoria, particularly in the goldfield districts. The exhaustion of the mixed species Wombat Forest by the end of the century (it was declared 'a ruined forest') turned the eyes of the sawmillers eastward and into the mountains.

The discovery of gold had decentralised the demand for timber. The timber tramway era, by contrast, was a consequence of a strengthening central market and the development of a means of reaching it. Melbourne's rapid expansion and feverish building in the 1880s created a flourishing, centralised timber market just at the time Victoria's railway system was extending its tentacles into the city's nearby forests. Trains first passed through Wandong and Nar Nar Goon in the 1870s, and arrived at Yea in 1883, Healesville in 1889, Fern Tree Gully in 1890, Gembrook in 1898, Warburton in 1901, Alexandra in 1909, Erica (and Walhalla) in 1910, and Noojee in 1919. These towns developed as railheads for the burgeoning mountain sawmilling industry.

Early sawmilling was concentrated in the slower growing, more durable eucalyptus species of the foothill forests. Utilisation of ash was considered unpayable due to the rough, high terrain it occupied. Bullocks and horses, which were used to haul timber out of the foothill forests, found the mountain ranges much harder going, and good feed was scarcer. The introduction of power logging – the use of steam-powered winches – made the steep ash forests millable.[9] And power logging could only be justified where a very high proportion of the trees were merchantable and there was therefore a large volume of timber per hectare to be removed, as was the case in the mostly even-aged, single-species ash forests.

However, there was also a problem with the timber itself that had to be overcome before large-scale utilisation of ash was profitable. Although ash species split well, they warped and collapsed when sawn and dried. From the 1870s to the early 1900s, sawmilling in the Mount Disappointment forest (where mountain ash is prevalent) concentrated on messmate logs instead. Messmate was also sought in the foothills of the Upper Yarra forest when it was opened up in the early 1900s.[10] It was not until kiln-drying and steam-seasoning techniques were successfully developed in the 1920s that the ash forests could be fully utilised. Steam heating (or 'reconditioning') of the partially dried timber restored the original shape and removed all stresses. By 1931 the increased use of local kiln-seasoned hardwoods was noted in the Forests Commission's annual report: '[I]t is estimated that 80 per cent of the flooring laid down in Melbourne during the past 12 months consists of kiln dried Victorian Mountain Ash.' On the eve of Black Friday, *Eucalyptus regnans* was acknowledged as 'probably the most important forest tree' in Victoria.[11]

These two technological advances – power logging and the reconditioning process – meant that the ash forests were logged intensively from the 1920s. Their very high timber yield allowed a higher density of sawmills than had been the case in the mostly messmate districts of Mount Disappointment and Kinglake, and the new technology demanded it.[12] The use of steam winches and other expensive and immobile machinery in mountain country meant that logging operations had to be concentrated and exhaustive. Whereas the mixed-species foothill forests – with trees of many sizes and ages, many of which had been fire damaged – had been selectively milled, the ash forests were clear-felled. The justification for clear-felling was primarily economic, but increasingly became silvicultural (that is, concerned with forest regeneration).

On the basis that sawn timber was cheaper and easier to transport than giant logs, mills were established deep in the forest, as close as possible to the areas of operations. A sawmill generally occupied one of the lower points in a logging area so that gravity would assist the transport of logs to the saw benches. A nearby source of water was essential to supply the steam boilers, lubricate the saws and for domestic use. Sawdust was left in large heaps, or found its way into streams or was burnt, sometimes to drive the boilers. The area around a sawmill was regularly burnt to maintain a clearing, to provide grass for animals and as protection from bushfire. Mills were often located on the rim of the mountain ash belt because, in the words of sawmiller Jack Ezard, 'to burn the mountain ash country is a very ticklish job'.[13] You either kill it or it kills you. And the timber was flawed for building in its untreated form. Sawmills were fashioned from their own forests and green mountain ash

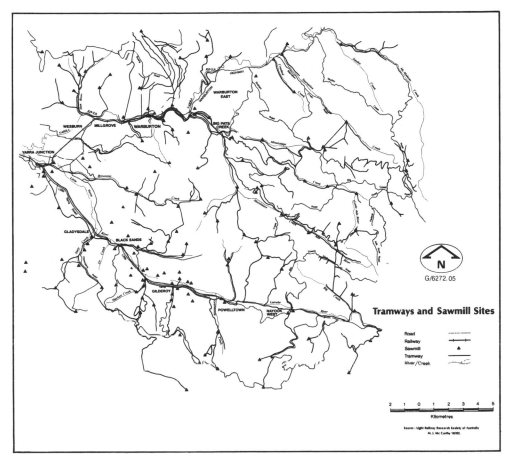

A tracery of tramways: timber tramways created an intense pattern of transport and settlement in areas now largely uninhabited. This map of the Warburton–Powelltown area is based on one drawn by M J McCarthy for the Light Railway Research Society of Australia. (DNRE)

created warped and draughty huts in country that was already cold enough. Also, ash in any form was useless for foundation timber, as it rots easily in the ground.

Tramways developed as a cheap, efficient and year-round means of conveying timber to the Victorian Railways lines, where it could then be made available to the Melbourne market. Although 'the timber tramway era' did develop this common pattern, the technology constantly evolved. The first lines were made with wooden rails and closely packed wooden sleepers that provided a firm base for horses or bullocks as they hauled the 'bogies' (trucks). Steel rails, sometimes used on corners, eventually totally replaced many wooden lines, and steam locomotives operated along

them. Rails on the timber tramways often had a colourful past, and some had earlier
served Melbourne's cable trams or Sorrento's steam and horse trams.

The high-yield ash forests made investment and innovation more likely and the
mountain terrain made it more necessary. Steam winches were introduced on inclines,
and impressive aerial cable or skyline systems dragged or lifted logs across valleys.
From 1934, caterpillar tractors surprised the old-timers with their ability to tackle the
steep slopes in forest haulage work. In the forest sawmills, the major technological
advances during this period were the introduction of the twin-circular breaking-
down saw bench, which could efficiently handle even the bigger logs, and the friction
powered roller saw bench. Lyrebirds learnt to imitate the entire workings of a bush
mill, even the distinct pitches of different saws.

Timber workers and their families mostly lived at the isolated sawmills. Big
sawmill settlements might have a tennis court, school or even their own football
team. At one old sawmill site, the net posts of the tennis court can still be found
beneath the scrub and blackberries.[14] For the women especially, a trip from the forest
sawmill to the nearest town might be a very occasional treat. The timber tramways,
along which some might travel only a few times a year, linked them to the world.

Even less isolated timber towns were socially dominated by the industry. Powelltown, in the Little Yarra Valley, was the headquarters of the Victorian Hardwood
Company, which from 1913 operated the only timber tramway licensed to carry
passengers (although many others did too). Although linked to Yarra Junction by this
excellent line, Powelltown remained little more than a sawmill settlement and a
company town. None of the town's residents could be said to be property owners. The
Forests Commission owned the area, surveyed it, laid out the streets and subdivided
it into allotments that were leased out and renewed annually. Most were leased not to
individuals but to their employer, the Victorian Hardwood Company. It was, officially,
a 'forest village'.[15]

Warburton, situated on the Upper Yarra River, was more than a timber town,
but it still hummed with the rhythms of sawmilling. In the words of one inter-
war resident, Alex Larkins, its exports were 'timber, tourism, water, biscuits and
literature'.[16] Many timber workers commuted weekly from Warburton to the forest
sawmills, but some bush workers were seen in town only at Christmas. Roughly
scrawled notes from the bush were answered with consignments of goods sent by
tramway.

Warburton's general stores supplied all the necessities: 'Varieties of tinned food, jelly crystals, tough clothes, plug tobacco, patent medicines, billies, frypans, heavy working boots and plenty of credit.' In the bush in the 1920s the working week was 48 hours (including Saturday morning), and whistle blasts around the hills announced the 7.30 am start and 5.15 pm finish.[17]

One child who grew up at a timber settlement near Big Pats Creek in the Warburton forest recalled that his home was made of rough timber, lined with hessian and covered with paper. Even the chimney was made of wood. 'Every night the fire had to be put out, followed by a good look up the chimney to see that it wasn't on fire.' He and other children at Mississippi Stables walked to school at Big Pats Creek along the tramway. 'In winter we used to get two long sticks and all hold on to them for pistons; and we would be a train and run all the way to school.'

He 'started in the timber' at the age of 15 and soon had one of the lowliest jobs in the sawmill: wheeling sawdust. Later he got to work out in the forest and learnt to get up on the boards above the tree butt, cut a scarf and then take his turn on the crosscut saw. Cutting a scarf – some called it 'putting the address on the tree' – was a precision job for it determined the direction in which the tree would fall. 'If you were good at falling you could put that tree just where you wanted it.' At the end of the day they made their own entertainment: '[W]e'd kick the football around the stumps at the mill.'[18]

The forests of ash were inspiration to C J Dennis, author of *The Songs of a Sentimental Bloke* (1915). He did much of his writing while living in the forest and was a joint proprietor of the Mount St Leonards Sawmill Company from 1918 to 1924. From 1908 until his death in 1938 he lived at Toolangi, first in the tent of artist Hal Waugh, then in an abandoned timber cutter's shack, and finally in a house called 'Arden' with his wife, Margaret Herron. He finished writing *The Sentimental Bloke* in an old tram on the property of J C Roberts at Sassafras in the Dandenongs. He witnessed and survived Black Sunday, the fire of 1926, and died half a year before Black Friday, 1939.[19]

In the forests of the Britannia Creek valley, between Warburton and Powelltown, was an unusual forest industry. From 1907 to 1924 Cuming Smith and Company operated a wood distillation plant that put to use those parts of the tree generally wasted in sawmilling. Mountain ash was the preferred timber. The best timber was sawn for sale, but lower grades were cut into 3 foot (0.9 metre) lengths and split and stacked for drying for over 18 months. The billets of wood were placed in wrought iron buggies and, after final drying in brick chambers, wheeled into retorts. The wood was carbonised, and the liquid and gas products were drawn off and treated. Methanol and acetic acid – raw materials in the chemical industry – were the chief products, and the residual charcoal was kept as valuable fuel. The industry employed 100 men and had a monthly payroll of £1000.

This 'Chemical Industry in a Victorian Forest', as the publicity literature put it, had its own school from 1912 and fielded its own football and cricket teams. The winding down of the industry affected form on the field: 'Britannia Creek were unfortunate in losing men just as they had a decent team together' commented a local reporter in 1923.[20]

In the Rubicon forest, tramways facilitated the construction of a hydroelectric scheme by the State Electricity Commission of Victoria. A scheme was first mooted for Snobs Creek as early as 1910, construction work commenced in late 1924 and five generating stations were opened in 1928. These were situated on tributaries of the upper Goulburn River and fed into a substation at Sugarloaf near the new Eildon dam, which had been created as part of an irrigation system for the lower Goulburn valley. The Eildon reservoir went through a further transformation for irrigation and power when it was greatly enlarged in the 1950s. The State Electricity Commission shared the Rubicon forest with the timber-getters, used their tramways to transport some of their equipment, and later perpetuated the art of trestle bridge building and maintenance on their own network of tramways.[21]

The saying goes that 'you always know an old sawmiller by his handshake'. It was a dangerous industry. The transparent spinning teeth of the saw blades swiftly took fingers or could spit lethal shafts of wood. Giant logs rolled unexpectedly, timber bogies hurtled into mountain gullies, and 'widow makers' – the high, dead limbs of trees – lived up to their name. Half the boot repairs in Warragul were for axe strokes. Brian O'Toole, who salvaged mountain ash killed in the Black Friday fires, had a finger chopped off, his hands torn by a chainsaw, an eye blinded by a flying chunk of steel from a wedge, and a leg crushed by a log. He was working among the tall ash when his 21-year-old brother Martin was killed instantly by a falling branch. 'I couldn't go back there working', he recalled. 'I didn't go into the bush for a while'.[22]

The skyline leads of the ash country, which were strung between lopped spar trees, were created by breathtaking feats of courage. High climbers donned spurs and belt, threw a rope around the trunk to give them grip, took a finely honed axe and 'walked up' the trees to lop the tops. Hec Ingram remembered the adrenalin, the butterflies and the wobbly knees that accompanied the moment of 'blast off' when 'the big trunk comes back and back, and in recoil literally launches the head into space'. The climber braced himself against the tree as the top fell and the trunk recoiled. In a gust of wind, the top could easily jerk back on the climber, or a split trunk could pin him, then break the rope that was his lifeline.[23]

Hec Ingram was born in 1909 on the western coast of New Zealand where his father, Chris, operated Kauri sawmills. He grew up in Powelltown where his father was managing director of the Victorian Hardwood Company. He experienced the 1926 fire at Erica and was involved in the salvage of fire-killed ash after the 1939

A timber worker climbs a mountain ash with spurs on his boots and a rope for grip. With his axe (which can be seen dangling from his belt), he will lop off the top, leaving a tall spar tree for use in a high lead logging system.
(DNRE Library)

High climber Charlie Wall tops a fire-killed mountain ash at East Tanjil in March 1946. He can be seen bracing himself as the head of the tree takes off and the trunk recoils.
(DNRE Library)

fires. He retired to a North Box Hill home built exclusively of mountain ash. Ingram regarded Jack Ezard as 'the greatest hardwood sawmiller' he ever knew. 'He was an innovative engineering and transport genius. No argument in my book.' Ezard was one of the pioneers of the skyline system of logging. Ingram and Ezard, along with Paul Christensen and the Saxton and Collins families, were among the great names of mountain ash sawmilling.[24]

Between two great Yarra River floods in 1891 and 1934, and between two economic depressions in the same decades, the Upper Yarra forests were sites of productivity and prosperity. The Upper Yarra Forest District, based at Powelltown, was the premier sawmilling region of Victoria. Sawmilling activity peaked in the 1920s. At the Powelltown mill at this time, the average daily production was over 37 000 super feet (over 87 cubic metres) – where a 'superficial foot' was a square foot of timber one inch thick, and a fair day's work by two men pit-sawing would produce 100 super feet (over 0.23 cubic metres). Yarra Junction, where the lines from the Warburton and Powelltown forests met, boasted that more timber passed through it than any other place in the world except Seattle in the United States. In the 1920s sometimes in excess of one million super feet (over 2300 cubic metres) of timber were air-dried there at any one time, awaiting despatch to Melbourne.[25]

In the same decade the sawmilling industry was taking off further east. As gold-mining declined in the Walhalla area, Erica took over as the district's major centre, and timber became a substantial source of employment. Mills sprang up in the forests from Moondarra, Erica and the Thomson Valley across the ranges to the Upper Tanjil Valley and Mount Baw Baw. By the early 1930s, sawmillers were entering the bush country near Matlock and on the Tanjil.

Four events finished off the Upper Yarra sawmilling heyday: a strike, a depression, a flood and a fire. Each seemed to come from elsewhere, from somewhere outside control; yet there was a nagging doubt that responsibility lay closer to home. In 1929, a strike of timber workers silenced the sawmills for months. This bitter industrial fight was in response to an arbitration award (the Lukin award) that increased weekly hours from 44 to 48 for city timber workers, sharply reduced wages and margins for skill, doubled the proportion of youths allowed in the industry, and permitted the introduction of piecework with no guaranteed base rate. The bush workers were already doing 48 hours a week but joined the ultimately unsuccessful strike for nearly five months.[26] The general economic depression followed. Many timber companies collapsed. Men were out of work and hundreds of draught horses were destroyed.[27] Output from the Upper Yarra Forest District plummeted from 3 217 822 super feet of sawn timber (over 7500 cubic metres) in October 1929 to 530 073 (over 1200 cubic metres) in January 1931.[28]

Relief workers were part of the Powelltown scene for several years. Even rabbiting wasn't much good in the dense ash forests where little grass grew, and men had better luck in the farming country around Nayook and Neerim.[29] As late as 1936, 200 relief employees still worked for the Forests Commission in the Powelltown subdistrict. The commission also established another form of unemployment relief; boys' camps where jobless youths were trained for forestry work. Throughout the 1930s, the virgin forests that had sustained the local economy were close to being 'cut out'.[30]

In 1934, as the Upper Yarra Forest District began to recover from the effects of depression, there was another setback: the Yarra River burst its banks in a very dramatic fashion. What would have been disruptive in any case was made a disaster by the events of previous years. By that year the banks of the Yarra from McVeighs to Warburton were littered with logs, trees and heads of trees left from half a century of sawmilling. As well, the river flats were covered with acres of stacked timber, the surplus from mills that became idle in 1929. In the hope of a new building boom the timber companies had kept these assets intact. Just as the light of recovery glimmered, the Yarra surged through the half-century's litter and the stacks of sawn timber planks and swept them violently downstream. Waste and assets together became the battering rams with which the river destroyed homes and bridges along its length.[31]

In the forests the discarded tree crowns lay unburnt and unused around the bush mills. Sawmillers neglected to build dugouts. In their annual report for 1938 the officers of the Forests Commission of Victoria reported that their fire protection organisation was proving effective. They had not lived long enough.

THE GOOD OIL

GARY PRESLAND

Sometimes we cannot see the wood for the trees. In the collections of Museum Victoria are samples of material wrought from the wood of the ash forests – oils, acetates, alcohols, and samples of the woods themselves. The existence of such items within the Museum's holdings reveals another industrial history of these forests and also something of the function of museums.

In the first decades of the twentieth century these forests were the subject of intense interest to companies looking to extract marketable products. From August 1907 until August 1924, one such company, Cuming Smith, operated a wood distillation plant at Britannia Creek, south of Warburton. At the peak of its operation the plant employed between 60 and 100 men in the process of distilling about 20 000 tons of wood per year.

Many trees had been felled for other purposes but not all the wood was suitable for building materials, palings, posts, or rails. This otherwise useless timber was used by Cuming Smith to produce a range of natural chemicals of potential economic value. The process of wood distillation allowed little waste – all parts of the trees were used, even the heartwood which was burnt in the furnaces to fire the process of distillation.

Among the materials derived by the Cuming Smith Company, Museum Victoria holds samples of charcoal, wood tar, tar oil, lime, methyl alcohol, acetate of lime liquor, formalin, calcium acetate, and acetic acid. All of these substances are produced in the process of dry distillation from *Eucalyptus regnans*.

Coincidentally, two of these products – methyl alcohol and its derivative formalin – have a practical application in the preparation of museum collections. Both these organic compounds can be used as agents to fix the molecular structure of animals when first collected.

The Museum's History and Technology collection also holds many samples of timbers of local origin, presented as exhibition pieces to showcase the natural assets of the Dandenong Ranges. One such piece, possibly of *E. regnans*, was prepared for exhibition in the Industrial and Technological Museum in 1885. The sample was subsequently exhibited to the public on a number of occasions: at the 1886 Indian and Colonial Exhibition, in London; at an 1887 Exhibition in Adelaide to mark the Jubilee of Queen Victoria's coronation; in Melbourne's 1888–89 Centennial Exhibition; and in the 1889 Exposition Universelle in Paris.

Cuming Smith's No. 2 sawmill at Britannia Creek, looking East. (Museum Victoria)

Another wooden showpiece also held by Museum Victoria is a veneered sideboard made from figured *E. regnans*. This was designed by P C J Glass of the Victorian Public Works Department and made by the Goldman Manufacturing Company in Melbourne. This sideboard was one of a series of articles that formed a furniture exhibit from the Victorian Government awarded a gold medal at the Panama Pacific International Exposition, San Francisco, 1915.

Such pieces serve as small but significant reminders of changing views of the forest and also the role of museums in showcasing resources and technologies.

7

WATER

On Black Friday large swathes of forested land in Melbourne's water catchments were severely burnt. The relationship between ash, fire and water was forged into public policy through the 1939 Royal Commission. 'To endanger the Mountain Ash forests on those watersheds is to actually endanger the supply of water', explained the engineer of water supply, A E Kelso.[1] He argued at the 1939 Royal Commission that no major fires had begun inside the catchments and that this was the pattern historically. '[I]n the history of the Board [of Works]', explained Kelso, 'the only fires which have come into our watersheds and have done damage – and they are very few – have been fires which had achieved tremendous dimensions before we could get at them.' This was a sideswipe at the Forests Commission, which managed much of the surrounding land. Fires that began in the catchment areas, Kelso explained, were generally put out swiftly and effectively. 'We have that advantage', he added wryly, 'that there is generally water available somewhere.'[2]

Almost all of Melbourne's water comes from the mountains of the ash forests. The city obtains about 90 per cent of its average annual water supply from these upland catchments, and 10 per cent from the Winneke scheme which fully treats the waters of the lower Yarra.[3] The earliest reservoir, Yan Yean, lies on the western fringe of the forests of ash, while the latest, the Thomson, is on the eastern edge. The use of this land for water catchment purposes has two major legacies: features of hydraulic engineering which date from as early as the 1850s, and large tracts of protected forest and flooded valleys which have a special status as public land.

The levels of dissolved material in these mountain streams are naturally very low because the landscape has been heavily eroded over long periods of geological time. The rivers run pure over old, worn rock. Although south-eastern Australia is poor in freshwater native fish, the invertebrate population of the streams is extremely rich. Our awareness of this diversity, and of the effects of changes in the forest upon this freshwater fauna, is quite recent, as it is only since the 1970s that there have been systematic surveys of aquatic ecosystems in Australia, and only since the 1980s that Museum Victoria has begun to build up reference collections of these less conspicuous insects, which were often overlooked by earlier naturalists.

Richard Marchant, freshwater ecologist at the Museum, emphasises the role of museums in recording environmental change and the value of scientific collections as resources that can be re-interpreted by future generations. The variety of caddis flies, stone flies, may flies, beetles and other life in the mountain streams is a useful signal of water quality. Also, such freshwater fauna does not easily disperse and is of ancient, Gondwanan origin, well-differentiated before 'the Great Upheaval', and so constitutes a clear biogeographic signature in the fossil record.

Museum Victoria and Monash University now regularly monitor the Acheron River which is typical of the cool (never much above 17°C), shaded, well-oxygenated streams of the ash forests, always overhung with sources of food supply in the forms of leaves, bark and other organic matter. Aquatic fungi and bacteria decompose this debris and then themselves become food for insect larvae and other invertebrates, thereby helping to infuse the cool, running water with nutrients and energy flows from the embowering forest. The health of vegetation and water are interdependent.[4]

Melbourne Water controls some 140 000 hectares of catchments, and most of that area is forested. Ash-type species (mainly mountain ash) form the dominant forest cover on 50 per cent of the catchment, and from that half of the catchment – the ash forests – comes 80 per cent of the average annual streamflow. About 20 per cent of the ash-type forest is over 150 years old, and the remainder is regrowth originating mainly from the 1939 fires.[5]

The East Plenty River, which rises in one of Victoria's tallest and oldest mountain ash forests, was the first organised source of Melbourne's water supply. But during its first two decades, Melbourne's population did not benefit from this pure mountain water, relying instead on tanks, wells and water from the Yarra. In August 1851, on the eve of a massive influx of gold-seekers, civil engineer James Blackburn proposed

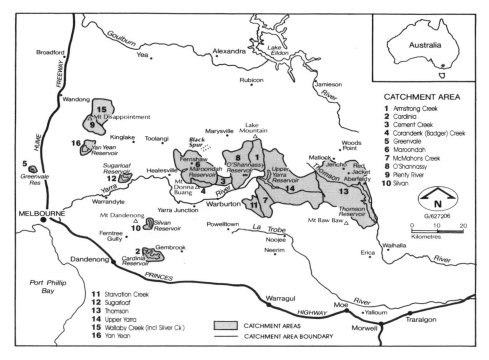

Melbourne's water supply catchments. (DNRE)

the construction of a reservoir at Yan Yean that would tap the Plenty River, which drained the granitic plateau of Mt Disappointment (the Hume Plateau). It was the colony's first major waterworks. A nine-metre high earth embankment was constructed to create a 29 million cubic metre reservoir from which a large iron pipe took water to Melbourne.

Water first reached the city on 31 December 1857, and from that moment, its quality was closely scrutinised by consumer and scientist alike. Early lead pipes poisoned the water, and high levels of organic matter made it unattractive and at times undrinkable. In assessing these problems, Melbourne scientists generated one of the earliest public discussions about Victoria's geology, hydrology and climate, and some emphasised the importance of home-grown environmental expertise.[6]

Yan Yean also attracted attention in the 1850s as one of the world's largest artificial water-storage reservoirs. It still remains part of the Melbourne water-supply system, although the early concern about the quality of its catchment area persisted and water for the reservoir was increasingly drawn from more distant sources. By August 1882, almost 1300 men were excavating channels, quarrying granite, shaping it into blocks and lining aqueducts, weirs and tunnels on the fern-covered flanks of Mount Disappointment.[7]

Construction of the Wallaby Creek weir and aqueduct in 1883 and the Silver Creek weir and aqueduct in 1886 drew water from streams north of the Great Divide along elegant channels over a low saddle east of Mt Disappointment. This remains the only part of the Melbourne system to draw water from north of the Great Divide, which is traditionally claimed for rural purposes rather than for the thirsty city. In the same decade, the building of the Clear Water Channel and the Toorourrong reservoir, and the exclusion of water from the western branch of the Plenty River and Scrubby Creek, enhanced the quality of the Yan Yean supply. For the first time all Melbourne's drinking water came from closed, unoccupied catchments.[8] The influx of mountain water even eliminated the algae which had bloomed discouragingly on the reservoir's surface. These dams, weirs and stone aqueducts are remarkable early feats of surveying and engineering and many of them, now notable historic features, continue in use.

A view along the Wallaby Creek aqueduct about 1905. The channel diverts the waters of Wallaby Creek from the River Murray watershed into the Plenty River or Yarra watershed. (Reproduced from MMBW, Water Supply, Sewerage, etc. Photographic Views, Osboldstone & Attkins, Melbourne, 1905.)

Melbourne's population grew quickly in the boom decade of the 1880s and settled increasingly in higher, eastern suburbs where Yan Yean water arrived at low pressure. Between 1886 and 1891, a weir was built across the Watts River (later known as Maroondah) and diverted by means of a 65-kilometre aqueduct to the Preston reservoir, built in 1864 for storage purposes in the Yan Yean system. To supplement this new supply a weir was built on the Graceburn River.

The Melbourne and Metropolitan Board of Works (MMBW, later known as Melbourne Water) was constituted in 1891 and embarked on its principal task of sewering 'Marvellous Smellboom', as the Sydney *Bulletin* labelled the burgeoning and filthy city.[9] For the next two decades, construction work for Melbourne's water supply was therefore minor: further diversions into the Watts River, the building of service reservoirs and extension of the mains to outer suburbs. Then from 1911 to 1914 a weir was built on the O'Shannassy River, and, a decade later, a dam. In the 1920s the Maroondah dam was also built, as was the Silvan reservoir in the Dandenongs.

This period of construction work was a response not only to Melbourne's rising population but also to a steady increase in average daily water consumption. In 1891 it was 240 litres per head; by 1940 it was 350 litres. (By 1990, it had reached almost 500 litres per head.)[10] People took Melbourne's role as capital of the Garden State seriously and poured enormous amounts of water into their front and back yards on hot summer evenings.[11]

The Upper Yarra dam, built upstream of Warburton, tripled the amount of water impounded for Melbourne's use.[12] Although its site was first determined in 1940, it was not completed until 1957. At its peak, 1300 workers were employed on the construction of the Upper Yarra dam, and a temporary township with a population of 1000 was established in the ranges beyond Warburton.[13] A severe drought ten years later provided further impetus to the board's building program, and new reservoirs at Cardinia Creek (filling 1973–77) and the Thomson Valley (dam wall completed 1983) each created new capacity records.

Every dam built and every tract of the ash forests claimed for water supply purposes plunged the Board of Works into prolonged political manoeuvring. Two major ongoing tensions entangled its efforts to keep up with Melbourne's growing population and even faster-escalating water consumption. One was the battle between the city and the country over the ownership of water resources, strung appropriately along 'the Great Divide'. The other was the fight with the Forests Commission over access to timber resources in water-catchment areas. Both of these tensions persist today and seem likely to erupt again in future planning decisions concerning the ash forests. At the heart of these debates is Melbourne Water's controversial closed catchment policy. The policy's evolving rationale has reflected changing scientific and political priorities in catchment management.

The policy of strict catchment control was well-established before the formation of the board in 1891 and was shaped by concerns about Yan Yean water quality and catchment access. The emerging science of bacteriology and beliefs about the positive relationship between forest cover, rainfall and streamflow set the context for the early debates.[14]

When the Yan Yean system was inaugurated in 1857, much of the catchment was inhabited farmland or forest plundered by paling splitters. These human intrusions dirtied and polluted Melbourne's drinking water, and in 1872 the remaining un-alienated land in the Yan Yean catchment and in the headwaters of the Plenty River was reserved for water-supply purposes. The role of water supply as a carrier of typhoid was shown in 1880, and death rates for the disease in Melbourne from 1870 to 1890 were four times those of similar British provincial towns.[15] Action commenced in the 1880s to remove all habitation and farming from the catchment, and the last Yan Yean property was bought in 1896. The board pointed out in 1942 that workers in catchment areas were 'blood tested to ensure they are not carriers of any water borne diseases and operate under the most rigid sanitary conditions'.[16]

Perceptions of the use and abuse of forests have also shaped the closed catchment policy. The belief that trees brought rain influenced early attitudes to the activities of paling splitters in the upper watershed of the Plenty River. In 1873, with the aim of retaining rainfall, the original 1872 reservation boundaries of the catchment supplying Yan Yean were extended over three kilometres north beyond the crest of the range, to limit the activities of the loggers and splitters. In 1879 all cutting, split-ting and carting of trees in the water supply reserve were prohibited.[17]

In 1872 the Watts River (Maroondah) catchment was recommended for reser-vation, and in 1888 the Upper Yarra and O'Shannassy catchment areas were permanently reserved. Tracks in the Watts River catchment area were kept open to the public as one of the conditions of board control. The township of Fernshaw, situated in the Watts River catchment, was bought out by the then Water Supply Department in 1885; the buildings were sold or demolished, and cesspits and other wastes carefully removed. Such actions did not please decentralists, who were urging the value of country settlement while, in their words, local land was being 'locked up' for city use.

One plan in 1901 to further augment the Yan Yean supply from the northward-flowing Acheron River encountered this strain of rural politics. The Forests Commission, together with country shires and water trusts, opposed the board's scheme on the ground that northern-flowing rivers were properly for the use of settlers on the dry northern plains and were not to be turned around for Melbourne's convenience. This principle received its baldest restatement in 1964 when a Victorian Premier with a strong rural bias, Henry Bolte, pre-empted discussions about diversion of the Big River by promising (on the eve of an election) that not 'one drop' of water would be diverted from north of the ranges to the metropolitan area.[18]

There was increasing reluctance on the part of the government to vest land outright in the board because of the effect on timber resources in the catchments. Although the O'Shannassy and Upper Yarra catchments were reserved in 1888, it was a further 22 years before O'Shannassy was vested in the board, and the Upper Yarra never was, although the board was given exclusive access to part of it in 1928. Two later catchments negotiated in the 1960s (the Yarra Tributaries and Thomson catchments) gave the board only shared control.

The disagreement between the Forests Commission and the Board of Works over the use of timber in the catchment areas has been a recurrent feature of policy debates about the ash forests. Theirs has been an argument partly about economics and institutional power, partly about the impact of timber harvesting on water quality and partly about what constitutes a healthy forest. These recurrent issues were studied in detail in 1957–60 when, in response to a deputation from rural bodies and the timber industry, the State Development Committee (SDC) inquired into whether water catchment areas should be opened up for timber utilisation. The committee recommended that controlled logging be allowed in the water supply catchments, but extensive public controversy led the government to make no change to the existing catchment policy.[19] Today, some small-scale logging is allowed, under strict controls, in formerly closed catchments.

The Forests Commission and sawmilling communities abhorred the locking up close to Melbourne of 'prime forest producing land' that could provide top-quality ash timber and employment for thousands. Their desire for access to the catchment forests was sharpened by the effects of the 1939 fire and by the wartime and post-war demand for timber. Sawmiller Hec Ingram confessed that 'it is sad to see, as I do every week, these majestic trees living out their life span in waste.'[20]

In some areas, reservation for water catchment entailed a direct usurpment of bureaucratic power. Why should the Forests Commission give up its management of some of the Upper Yarra forests, which increasingly needed reafforestation, to a body that employed no professional forester until 1940?[21] Foresters considered the board neglectful of its forests. It was not just a question of waste; a healthy forest was a used and managed forest. In 1895 the visiting Indian forester Inspector-General Ribbentrop was shocked by what he regarded as the waste of timber in the board's catchments: '[T]he ground is covered with dead trees and other vegetable debris in all stages of decomposition, and other dead and dying trees are ready to replace the fallen giants as fire and rot consumes them'.[22] He believed that these forests should be systematically worked.

The debate about whether timber-harvesting is compatible with water conservation began in the 1860s and continues today. Foresters continually argued that sensitive access to catchment forests would not upset water quality. The board replied with arguments about the risk of biological contamination and the increased danger of fire. Fire destroyed the leaf litter and mulch that helped retain water run-off, increased erosion and siltation, and threatened the mountain ash that grew in the most productive parts of the catchment. For these reasons the board did not conduct fuel reduction burning in its forests. It aimed for the total exclusion of fire.

A regenerating forest, with fast-growing young trees, uses more water than a mature one and leaves less for run-off. By the 1920s the defence of the closed catchment policy rested not only on the public health issues that had shaped it, but increasingly on the greater efficiency of undisturbed forests as water collectors. It was a question of water yield as well as water quality. On Black Friday about two-thirds of the Maroondah and large sections of the Upper Yarra catchments were severely burnt, as was the Silvan Dam. The O'Shannassy was only partly damaged, and Yan Yean and Wallaby Creek escaped almost entirely. Following 1939 and the growth of dense young forests the harvest of water steadily declined by an average of a quarter, regardless of rainfall figures.[23]

The board's first venture in catchment hydrology began in 1948 when it sponsored the investigations of a Master of Science student, John Brookes, into the water use of catchment vegetation at Wallaby Creek.[24] The run-off there had shown a marked decrease in the previous 20 years, possibly due to the 1926 fires that had invaded the northern and central parts of the plateau. Brookes sampled soil moisture in different-aged stands of ash and patches of bracken by measuring the electrical resistance of porous absorption units buried in the soil. He found that the young forests use the most water and bracken the least. He also quantified the ability of the mountain ash to intercept fog and low cloud and produce water run-off. Cold fronts that do not produce rain often carry cloud through the tree canopies at the height of 600–700 metres. In the ash forests, fog-drip adds significantly (15 to 20 per cent) to the total precipitation in some mature mountain ash forests.[25] Trees do bring 'rain' after all.

After Black Friday, the board invested more resources in the maintenance of the catchment forests, through the building of fire-breaks, some fuel-reduction burning in the surrounding buffer zones, and greater efforts at reafforestation to stop the mountain ash forests 'drifting back to bracken'.[26] A giant of a forester, 'Tiny' Bob Oldham, revolutionised forestry in the Board of Works when he became its first forest officer in 1940. He was among the first to work out that, contrary to popular belief, nitrogen fertiliser enabled mountain ash seedlings to become established in bracken, thereby saving the trouble of repeatedly clearing it.[27]

Research into catchment hydrology became politically pressing after the State Development Committee Report of 1960. In 1967 the board initiated further experiments at Coranderrk and North Maroondah to investigate whether, and under what conditions, controlled logging was feasible in catchments without there being adverse effects on water quality or quantity. Research indicated that logging operations and the construction of roads, if done under rigorous supervision, could be carried out in certain areas without diminishing water quality. However, in terms of water yield it was confirmed that fire and timber-harvesting have a major impact on run-off. Reductions in streamflow of up to 50 per cent (after an initial increase) had occurred following the 1939 fires, and it might take an ash forest 150 years to recover its pre-fire water yields.[28]

Water yield had clearly taken over from public health as the strongest argument in favour of a closed catchment policy. Both rationales were utilitarian and human-centred. They concerned the quality and availability of water for a city of consumers. The board aimed to maximise both factors. Tony Dingle and Carolyn Rasmussen, in their centenary history of the board, described how the organisation had little interest in persuading people to use less water: 'The more water it sold the greater its revenue in a supply system where capital costs were high but operating costs [due to minimal chemical treatment] were low.'[29]

Dam-building had been the solution to earlier droughts and water crises, but by the 1980s the building of dams was fraught with political difficulties following the rise of a powerful environmental lobby. Melbourne's Board of Works began to address the problem of escalating water consumption and to discourage the domestic user from being 'a Wally with water'.[30] In a television advertisement urging water conservation the board drew attention to an unflooded valley in the ash forests and pointed out that it would need to be dammed – unless city consumption was curbed.

At the same time, an entirely novel rationale for the board's closed catchment policy emerged – one that its founders, living in the age of acclimatisation societies, could not have foreseen. For over a century great tracts of natural habitat had been preserved, literally buffered from civilisation; so the catchment areas have nature conservation and 'wilderness' values of outstanding significance.

SHORTFIN EEL

MARTIN GOMON

Mention fish migrations and we generally think of salmon weaving their way through treacherously swollen streams and rivers fed by spring melt, leaping apparently insurmountable waterfalls and progressing toward their customary spawning sites. As remarkable as the trips of salmon are, their journey is no greater than that of the shortfin eel, *Anguilla australis*, or, for that matter, any of the other species in the freshwater eel family Anguillidae, spread around the world. Members of the family are the only true freshwater examples of what is otherwise a vast assemblage of serpentine marine fishes.

Shortfin eels pass their adult lives in the streams, rivers, billabongs and lakes at the upstream end of the tributaries flowing into Australia's south-eastern coast. After a number of years of relatively quiet existence feasting on the abundant animal life of these waters, sexually mature adults leave their sheltered homes and swim not merely down the rivers to estuaries, but much further afield, along the open coasts, against the influences of the powerful South-eastern Australian Current, to the Coral Sea thousands of kilometres away. Here they congregate, breed and die. Their eggs and then their subsequently hatched, leaf-like, transparent leptocephalus larvae are carried on a reverse journey of up to three years by the same prevailing current that resisted the efforts of the dedicated parents.

On reaching south coast waters, leptocephali metamorphose into tiny eel-like juveniles, or 'silver eels'. These diminutive versions of adults are capable swimmers and they make their ways to the estuaries and rivers that were the homes of their parents. Young eels arrive in the Gippsland region of eastern Victoria around the month of May but do not reach the western portions of the Victorian coast until October.

Upstream migrations, like the downstream journeys of adults, are not without obstacles, and on many an occasion after heavy rains eels have been found snaking their way across soaked paddocks in search of isolated ponds or a more direct route to the sea. Back in the headwaters of catchments, maturing fish assume the predatory lifestyles that characterise the species.

The regularity of the timing of these migrations was not overlooked by indigenous inhabitants of this region who incorporated it into their annual calendar by planning for the appearance of migrating eels. They timed their visits to riverine areas to

Apart from their dark eyes, the transparent marine 'leptocephalus' young of freshwater eels are well suited for life in clear, oceanic waters. (Shun Watenabe)

periodically fish for eels with beautifully woven, funnel-style eel traps that faced downstream.

The shortfin eel shares its lifestyle with about 15 closely related species living on continents around the world, and the breeding localities of several are in the same areas of major ocean basins. The repetitive pattern reflects their common ancestry.

8

THE THEATRE
OF NATURE

In November 1889 six members of the Field Naturalists' Club of Victoria, formed nine years before, set out in search of the Yarra Falls. One of the party was Melbourne University's newly appointed biology professor from England, Baldwin Spencer, who was keen for some Australian bush experience. He and his friends wanted to know something of the source of Melbourne's river and to see with their own eyes, and to capture on camera, the great Yarra Falls.[1]

In Marysville they were told that few had ever reached the elusive falls, although several exploring parties had made the attempt. Just as they set out, it started to rain. It rained for three days. At least they knew that if they reached the falls they would be thunderous. They followed an old surveyors' track cut in the 1860s, now grown over but still marked by the stumps of trees that had been felled. When they had to leave it, they used their compass and blazed the trees with a tomahawk. It was wet, hard going, stumbling over the roots of giant beeches, climbing over fallen mountain ash, brushing past luminous apricot-coloured fungi. Even in the rain, Baldwin Spencer could thrill at the sight of a myrtle beech forest whose warm brown tones reminded him of the English woods of home. The scrub, however, as he put it, 'always seems to grow just to the right height to soak your legs through and through, and to send showers of cold spray down your neck.' They became drenched. They walked in the moonlight. They cut scrub and tree-fern fronds to make a floor for their tent. Days passed. The food ran out. Nevertheless they pushed on and even collected on the way, shaking branches into upturned umbrellas and identifying three new types of worm.

The falls, which they did find, were worth it: the water cascaded violently. One of them cut a big shield in the bark of a tree on the bank of the river, with the letters of the club, FNC, and the date.[2] They took the first photographs of the Yarra Falls. Perhaps they were amused to find the slow, murky Yarra so enraged and white. In just over a year, in 1891, the Yarra would flood dramatically and bring that fury back to the heart of Melbourne.

Victoria's naturalists drew on the forests of ash for much of their sense of region and identity. If people from New South Wales were popularly known as 'Cornstalks', South Australians as 'Croweaters', Queenslanders as 'Bananalanders', and Western Australians as 'Sandgropers', then the distinguishing environmental fate of Victorians was that they were 'Gumsuckers'. Whether sucking sweet gum from the trunks of wattles, or simply growing up among the gum trees, colonial Victorians were identified as forest dwellers. And the great forest, this lush source of distinctive natural symbols, was close by in Melbourne's mountains. Both the state's faunal emblems, the Helmeted Honeyeater and Leadbeater's Possum, are only found in or near the forests of ash – and both are endangered. The Superb Lyrebird (*Menura novaehollandiae*) has a wide distribution throughout south-eastern Australia, but Melbourne's Sherbrooke Forest – on the southern face of the Dandenongs – became its most famous home. International visitors to the city were taken into the hills to glimpse the lyrebird's display and hear its mimicry, and to gaze up the white boles of the tallest ash. The Field Naturalists' Club of Victoria, made up mainly of urban, middle-class professionals, held many excursions into this region, systematically explored it, collected from it, and continually celebrated their own historical associations with it.

Investigating the Yarra Falls was to re-enact the excursion of Baldwin Spencer in 1891 and to fancy that one could recognise the same rocky wall down which that bedraggled party had clambered, or to follow in the footsteps of famed ornithologist A J Campbell who visited the Falls in December 1904, and to find his initials carved deeply into a sassafras tree.[3]

The collecting bags of expeditions were avidly compared for competitive and scientific purposes, the respective numbers of fern species calculated, and changes in the environment thoughtfully monitored. They were always looking out for 'our beautiful mountain butterfly', MacLeay's Swallowtail (*Graphium macleayanus*), probably the only naturally endemic permanent swallowtail butterfly in Victoria.[4] (But it had to be fit for the cabinet, for when a solitary specimen was taken in 1908, 'being faded and worn, [it] was set at liberty again'.[5]) A 'good haul of the river blackfish' might be caught for breakfast, and the fruits of indigenous plants nibbled tentatively and learnedly.

Near the end of September 1905, a small band of naturalists crossed the Blacks' Spur on foot in heavy snow, claiming that in their party was the first woman to do so in such conditions. They observed the effect of a foot of snow on the wildlife, even commenting on 'a solitary mosquito' which, when they attempted to capture it 'for our entomological friends ... flew away as airily and healthfully as though it were a summer evening'. An earthworm was regarded with awe as it crossed the road on deep snow, the sustaining ground so far beneath it. The 'Coachwhip-bird' kept whistling. No lyrebirds were seen, although they were said to be more numerous of late in spite of the foxes.[6] The Club recorded a corporate satisfaction at the achievements of first sightings, 'the first lady', the youngest person, the heaviest snow, or the biggest party to get most deeply into the forests.

Naturalists have often displayed what one historian has aptly called 'a mild theatrical streak', expressed in their prominent dress and technology – a characteristic of a marginal group seeking symbols of their identity.[7] They often relished and reported on their comic sense of difference. In January 1907 when a party of naturalists walked to the Yarra Falls, their arrival at the mining camp of Bromley's Reef on Contention Creek 'took the miners by surprise, and I have since learned that they thought a circus had arrived for their edification.'[8] Arthur Dendy patrolled the bush with his hunting equipment of one large and several small bottles of methylated spirits, a pair of steel forceps 'for the benefit of any beast which looked as if it might bite', and paper and pencil for sketching and describing the size, colour and form of his animals before the plunge into spirits shrivelled and dulled them.[9] He reported to the club on his quest for the worms, insects, centipedes and molluscs near Walhalla, these 'light-abhorring, lower forms of life', 'the fauna of dark and narrow crevices' which were laid open to him by the timber tramways snaking through the bush.

The tramways created convenient corridors of natural history enquiry, just as the surveyors' tracks and miners' paths exposed the bush to the magnifying glass. Henry Tisdall, who was a naturalist of note and a Fellow of the Linnean Society of London, as well as being the Walhalla schoolteacher for 18 years (1868–86), determined his collecting advice on the basis of the tramways and their aspects. When seeking out flatworms in the forests, he advised, look for 'the blue Planarian' (*Geoplana spenceri*) under large stones or logs on tramways facing the sun, 'the yellow Planarian' (*Geoplana mediolineata*) on shaded tramways facing the south, and 'the brown Planarian' (*Geoplana Walhalla*) on damp and deserted tramways facing south.[10] Tisdall also systematically searched the forests above Walhalla, on the eastern slopes of Mount Baw Baw, in search of fungi, seeking out favourite dells, turning over pieces of bark, lifting ferns in 'very deep, dark gulleys', and examining decayed logs.[11]

Avid naturalists spent their evenings on club excursions setting butterflies, skinning birds, and identifying plants collected. Nothing was let go (unless it be 'faded and worn'). Snakes were always energetically pursued, even into streams and aqueducts, fished out and killed, 'promptly despatched', their heads occasionally bagged. Sometimes the only sighted snake was one draped dead over a gatepost, the trophy of a preceding nature-lover. An echidna seen crossing a track near their camp 'quickly found its way into a box, in which it was forwarded to the Zoological Gardens'.[12] King parrots were observed to be 'very scarce', even (perhaps especially) at King Parrot Creek near Mount Disappointment in June 1907: 'It may be remarked that occasionally a well-plumaged specimen will bring at least a pound in the city market'.[13] And the farmer's view of wombats as pests was often shared by the naturalist: '[O]ne stupid looking animal was almost run over by the buggy, a flick from the driver's whip failing to hurry it in the least'.[14]

The first 'camping-out' excursion of the Field Naturalists' Club of Victoria was held on the edge of the forests of ash in November 1884, four years after the club's foundation. 'All departments of natural history were represented' at the gathering at Olinda Creek in the Dandenongs and the group included A J Campbell and the photographer Nicholas Caire. The first day '[b]eing Sunday the guns were left behind till the morrow'. But by the following morning, the party was so bristling with arms that new arrivals 'were at first afraid they had been conducted into the midst of a band of Italian banditti, so varied and formidable looking were the clothes and weapons of the party'.[15] The ornithologists were particularly triumphant 'in taking for the first time the nest and eggs of the rare and certainly the most beautiful of all the Australian honey-eaters', the Helmeted Honeyeater (*Lichenostomus melanops cassidix*), represented now by only about 70 individuals in the wild.[16] A J Campbell recalled the capture:

The nest was situated at a height of about twenty feet, and was suspended to
an outstretched branch of a hazel overhanging the creek. With what ecstasy
of delight the small tree was ascended! The handsome bird still retained
possession of its nest. With Mr Haddon's assistance, I all but had my hands
on the coveted prize, when, without a moment's warning, crash went the tree
by the root, and all – the two naturalists, tree, bird, nest, and eggs – went
headlong into the stream beneath. Alas! … But imagine our astonishment
when, after dragging ourselves out of the water, and removing some of the
fallen debris, we find nest and eggs intact – thanks to the poor bird, that
bravely stuck to its home till overwhelmed by the falling foliage. The eggs,
in which incubation had just commenced, were beautiful specimens, and
are now in my cabinet.[17]

At the end of 1898, a small party of ornithologists found a nest of Lewin's
honeyeater containing only feathers of the bird and a number of broken eggs, evi-
dently robbed by an owl or butcher bird, and proving 'that other creatures besides
field naturalists rob birds' nests'.[18]

A J Campbell also vividly recalled his first 'adventures amongst lyrebirds' in the
tall forests of the Neerim district, which was known as Buln Buln, the local Aboriginal
words for the bird's double call. He revelled in 'the threefold forest' of lower fern-
trees, medium-stature sassafras, blackwood and other acacias, and 'towering over all,
gigantic eucalytpts'. What a scene for the great hunt!

I leisurely ascended a gully or patrolled a recent survey line till I heard a bird
singing a little distance off in the scrub. Then I commenced very carefully –
for at a false move, an extra shuffle of the leaves, or the snapping of a twig
your prey disappears as if by magic – to crawl on my hands and knees, as
often as not wriggling snake fashion on my stomach, through ferns and
scrub, from tree to tree. The bird ceases singing, as if knowing intuitively
that danger is near. I stop, too, and pose like a stump. I dare not move a
muscle, although I feel a land leech attacking my ankles and a large forest
mosquito stinging the tip of my nose. Presently the bird commences whistling
as joyfully as ever. On I creep, every yard nearer, and with the excitement my
heart quickens its beat. It throbs so loudly that I fear the bird will hear. The

exertion in moist, thundery weather bathes me with perspiration; great beads roll off my forehead and patter down on the dried leaves. Affairs are now desperate, for at last I am within shooting range, and am peering through the ferns with gun uplifted and finger trembling upon the trigger ...[19]

The study of natural history and the culture of hunting were closely aligned in the nineteenth century. Hunting and collecting were therefore respectable and often synonymous. Naturalists were excited by the thrill of the chase and the identification and possession of a new specimen. They compared 'bags' and jealously guarded their 'hunting grounds'. Fern cases, butterfly cabinets, seaweed albums and egg and shell collections were common adornments of

The noted ornithologist, A J Campbell. In 1884, his party of naturalists were so bristling with arms that they were compared to 'a band of Italian banditti'. (Museum Victoria)

the drawing rooms of middle-class families. Collecting represented rational amusement, spiritual enlightenment and healthy recreation. Museums and libraries, then, were showcases of the prevailing ethic of progress, part of an extending imperial frontier, and evidence of the reclamation of the 'forest' by 'civilisation' and 'science'. The National Museum of Victoria, founded in 1854, energetically participated in this conversion, and a number of rare species from the tall forests were named after the museum's taxidermist.[20]

But in the early twentieth century, the forests of ash were also the stage for enacting changing sensibilities towards nature. Twenty years after the Field Naturalists' Club's first excursion of armed 'banditti' to Olinda Creek, three young nature-lovers – Charles Barrett, Claude Kinane and Brooke Nicholls – established camp at a bush hut in the same valley. Inspired by Henry Thoreau, they named their hut 'Walden'.

But, as Barrett put it, '[u]nlike our Master, we paid taxes, having no Emerson to pay them for us; and also, we welcomed visitors, which Thoreau seldom did.'[21]

Their plan was to spend all their weekends and holidays there and to monitor carefully, in words and photos, the natural history of their neighbourhood. Barrett wrote in 1905 that '[w]e desired to experience that return to Nature of which so much has been written in recent years; to leave the din and dust of the great city, and dwell awhile in the forest among birds and flowers and trees.'[22] Their hut, set in the midst of an old orchard near Olinda Creek and a pleasant walk from the Lilydale railway station, was a base for bird-watching, plant identification, photography, a daily notation of observations, and plenty of yarning and letter-writing.

Once ensconced in their chosen valley in the Dandenongs, the three nature-lovers were quick to compare it with the 'hill, dale, woodland and water' of Gilbert White's famous English parish of Selborne. They called themselves 'The Woodlanders', thereby invoking another recorder of rural ways, Thomas Hardy, whose novel of that name was published in 1887.

At this Australian 'Walden', the camera, like the pen, was wielded with a self-conscious, imitative air. While Barrett was treasuring associations with great nature writers such as Gilbert White, Henry Thoreau, John Burroughs and W H Hudson, Claude Kinane ('The Artist', i.e. photographer) was sending some of his snapshots across the seas to the pioneering British nature photographers, Cherry and Richard Kearton.[23] Their *British Birds' Nests*, published in 1895, was the first bird book illustrated with photographs all taken in the wild, and the Keartons' *With Nature and a Camera* had encouraged the 'Woodlanders' to try their bark hut experiment.

Local photographers such as J W Lindt and Nicholas Caire had, since 1880, been freed by the dry-plate process to venture into fern gullies and forests. But Kinane's intimate pictures of birds in their nests were some of the earliest such images to appear in Australia. Barrett had his own first glimpse of the Helmeted Honeyeater near the spot where that first nest was so dramatically 'taken' and in 1912 he took the 'pioneer' photographs of the bird, recalling 'I am not a collector, but delight in the beauty of birds' eggs; and looking for the first time upon a clutch of *M. cassidix* gave me the thrill of discovery – that of possession I did not desire.'[24]

The most famous photographs taken by the Woodlanders were those of a baby cuckoo, hours old, pushing its host chicks from the nest. These images were exciting achievements because they were action photos taken in the wild, but they were also morally challenging. Could such behaviour be natural? And could one simply observe it and do nothing? W H Hudson had considered cuckoos possessed of 'a devilish intelligence' and the Woodlanders repeatedly placed the abused chicks back in the nest, to no avail.[25] Yet their whole ethic, the very reason for their sojourn in the forests, was to enact a new relationship to nature, one of reverence and non-interference. Most naturalists of this period began with a shanghai in their pocket, and talked lovingly of the 'prizes', 'spoils' and 'trophies' won from their 'collecting grounds'.

The message of the Woodlanders was that nature could be captured with camera and field glass instead of with gun and trap. Barrett believed that 'the only way to study bird-life is to dwell among the feathered folk themselves', and he contrasted this method with that of the 'cabinet naturalists', satisfied only with their trophies.[26] But 'enlarged cabinets' remained a measure of success and an emblem of class among many field naturalists well into the twentieth century, and the Royal Australasian Ornithologists' Union (RAOU), founded in 1901, did not restrict the collection of bird skins and eggs among its members until the 1920s and '30s.[27]

Another piece of theatre was staged in the tall forests, quietly and devotedly, by Tom Tregellas (1864–1938) in the 1920s and '30s. And his co-conspirators in this opera were natural performers, the lyrebirds of Sherbrooke. Tregellas lived in the Melbourne suburb of Chatham but spent virtually all his spare time dwelling in a large hollow log in state forest near Kallista, 50 kilometres east of the city. Fire had already scoured out the interior of his home for him. The log was about five metres long and two and a half metres wide. He boarded up one end, built a canopy out of palings over the entrance and a fireplace in front, and called his home 'Menura'. He had renounced his earlier life as a collector (he had gathered specimens for Gregory Mathews who was preparing his *Birds*

Charles Stone, an enthusiastic member of the Royal Australasian Ornithologists' Union (founded in 1901), pictured with a baby lyrebird. (Museum Victoria)

*Tom Tregellas framed by the entrance to the
hollow log he made his home in order to study
the lyrebirds of Sherbrooke. (Photo by Michael
Sharland, reproduced from the* Australian Bird
Watcher, *vol. 9(4), December 1981.)*

of Australia), and aimed instead to secure the protection of the Lyrebird through his articles, photos and lantern-slide talks.[28]

Tregellas spent at least 18 years 'sojourning amongst them'. His knowledge, like that of the Woodlanders and many amateur naturalists, was proudly original. Originality was to be discovered through a direct, watchful, sensuous engagement with nature. 'I may also state at this juncture', he declared, 'that I have drawn on no one for my information, all particulars in regard to the birds being gleaned on the spot and all photos, absolutely my own'.[29] Through pedantic observation he corrected what he considered to be popular errors about his friends: lyrebirds do not dance, he insisted, they 'merely strut about and turn round and round', occasionally give a forward jump and 'take two steps backwards to the first position'. But although he dismissed the term 'dancing', he contributed to other aspects of the romantic cult of the Lyrebird, reporting their behaviour on 'state occasions', and describing their calls as 'amorous' or 'endearing'.

I have been with the birds at all times and in all circumstances. I have seen them in every imaginable attitude on mounds and open spaces; I have been with them through all winds and weathers, at their toilets and devotions, their displays and amours. I have witnessed the wooing and winning and been present at the marriage ceremony ...[30]

He wrote of intimacy, confidence and friendship. From being 'in company with the birds for so many years, they have become accustomed to my presence, and permit familiarities that make for our mutual benefit', he explained. The birds vouch-safed their secrets to him. Such a privileged status enabled him to offer 'The Truth About the Lyrebird', as he called one of his articles. His science was driven by the magic of proximity with beautiful, shy creatures rather than any systematic biology.

In 1931, Tregellas participated in broadcasting the song of the Lyrebird on ABC radio from the depths of the forest across the nation. Another ornithologist, Ray Littlejohns, helped make Sherbrooke famous with his filming of the lyrebirds there from 1925. In 1926, the Forests Commission determined that Sherbrooke Forest should be free from commercial exploitation and from fire.[31] Sherbrooke's two most famous lyrebirds, Timmy (1927–53) and Spotty (1942–64) became international celebrities. Each of the Lyrebird naturalists cultivated a personal relationship with the birds, giving them names (Silvertail, The Wanderer, The Broadcaster, Joe), observing them around the clock, analysing their rituals, and dramatising for a growing public their easy intimacy with such wondrous creatures.

Lyrebirds are large, have big, oval eyes, voices that can imitate human and artificial sounds, extravagant tails, mounds upon which they perform, and since the introduction of foxes, roost incongruously in tall trees. At Sherbrooke, they were available to a big city audience, were attributed with personalities, and – although apparently secretive and shy – actually abided humans and even developed long-term, faithful relationships with particular people. Not only did they have their own natural theatre, they seemed to be willing participants in a series of scripted dramas fulfilling the cultural need of Australians in the early twentieth century for a new, sentimental relationship with indigenous fauna.

When A H Chisholm, a noted Sydney naturalist, moved to Melbourne in the early 1930s, it was Sherbrooke's lyrebirds (and the Helmeted Honeyeater) that helped him reconcile himself to his new ornithological neighbourhood. Lyrebirds were, of course, very familiar to him and he had often 'delicately stalked' them in the Sydney region, grateful for a mere glimpse. In Sherbrooke Forest, he had 'never seen Lyre-birds to such advantage in any other part of Australia'. There they 'rollicked casually' into one's gaze and before one's camera lens and up to one's microphone.[32]

The growing silence of Sherbrooke has thus been particularly chastening to Victorians. Lyrebirds are found in many parts of Australia, and now also in Tasmania, where they were successfully introduced into a fox-free environment in 1934–49. They are common and widespread and are not endangered as a species, but in

HAIRSTREAK BUTTERFLY

ROSS FIELD

The tall forests of south-eastern Australia are rich in a wondrous variety of butterflies. Over 70 species have been recorded from the mountain ash forests and the subalpine meadows just above them. This rich fauna is dominated by a diversity of grass and sedge-feeding browns (Nymphalidae) and skippers (Hesperiidae), but it is the blues (Lycaenidae) that exemplify the intricate and delicately balanced nature of the forest ecosystems. The Silky Hairstreak (*Pseudalmenus chlorinda zephyrus*), a rare gregarious lycaenid that breeds on blackwood (*Acacia melanoxylon*), silver wattle (*A. dealbata*) and other wattles, has an extraordinary relationship with its food plants, colonies of small black ants, *Anonychomyrma biconvexa* and mountain ash.

This small, beautiful butterfly with delicate tails lays its eggs on wattle stems; but only on those that have *A. biconvexa* nearby. Usually these ants are attending sap-sucking bugs, milking them of honey dew. The ants soon focus their attention on the young butterfly caterpillars that hatch from the eggs. The ants swarm over the larvae, tending posterior glands that exude protein-rich sugars.

The ants tend the caterpillars throughout their growth, usually from November to January. When mature, the caterpillars seek shelter to pupate. This begins with an extraordinary journey. Down the wattle tree the mature caterpillars travel, following the trail of ants, and up and under the bark of an adjoining eucalyptus tree, more often than not a mountain ash. Here the caterpillars pupate, often clustered together, spending the next eight to ten months securely adhering to the trunk of the tree, and receiving occasional visits from the ants.

Such an amazing symbiotic relationship between ants and butterfly is thought to provide the caterpillars with some protection from marauding predators such as spiders and bugs and parasitic flies and wasps. At the same time, the ants benefit from the harvest of nutrient rich secretions. Although the ants are abundant in the forest, revealing their presence through a waft of formic acid when they are unduly disturbed, they do not need the butterfly to survive. The same is not the case for the butterfly. The Hairstreak cannot exist without attendant colonies of this one species of ant.

The Silky Hairstreak is one of the earliest butterflies to emerge in the spring and can sometimes be seen searching wattle foliage for suitable egg-laying sites or visiting nearby flowers to stock up on energy.

The Silky Hairstreak, Pseudalmenus chlorinda, *occurs in colonies, feeding on Silver Wattle and Blackwood in the tall forests.* (Ross Field)

Although widespread in its distribution it is not an abundant species. However, large colonies of the butterfly can occur and with careful searching it is not difficult to observe the larvae on the leaves of the wattles, the flurry of attendant ants betraying the location of the caterpillars. One such colony has existed for many years at Grant's Picnic Ground, Kallista, a popular tourist and barbecue destination for visitors to the Dandenong Ranges. Conservation measures are in place in the Dandenong Ranges National Park to protect this colony of the butterfly. Interpretative signs have also been installed in the picnic grounds to focus visitors on this intricate, usually unnoticed, natural relationship.

Sherbrooke they may face local extinction. How important is it to save them on this famous island amidst the encroachment of suburbia? If the ecological imperatives are debatable, what of the cultural significance of this site of intimacy with wildlife?

The predicament of the Sherbrooke lyrebirds raises important historical questions, both cultural and natural. The birds are not being loved to death, not quite – people in themselves are not a threat to the lyrebirds. The 'shy' lyrebirds of Sherbrooke have flattered many an ordinary observer with their social skills. L H Smith, another naturalist who has devoted the spare hours of a lifetime to cultivating intimacy with the birds, insists on the compatibility of birds and caring humans and reports as bizarre proof that 'there was one Lyrebird which took cheese held in my teeth'.[33] He goes on to argue that the fundamental reasons for the decline in the forest's lyrebird population are excessive accumulation of forest litter, invasion by weeds and noxious plants, and predation by foxes and cats. Wire grass (*Tetrarrhena juncea*) has grown into tangled masses over a metre high, closing off ground that was previously ideal for the scratching of lyrebirds which feed entirely from the leaf litter and soil. This tenacious grass was once held in check by wombats (another population that has declined) and by regular fire.[34] Suburbia has brought a fear of fire and also new predators; even in Tom Tregellas's time, lyrebirds had learned to imitate the yelp of foxes. In this particular forest, with this particular history, lyrebirds will need a rather different stage-management to survive.[35]

9

TOURISM

Tourism 'took off' in the 1880s as Melbourne's population grew rapidly and railways extended further into the hinterland. The wedge of forested land between the Sydney and Gippsland railways attracted visitors with its 'sweet-smelling fern gullies, fragrant bowers, giant towering trees, crystal streams and splashing cascades'.[1] One man's excursion into 'the Black Spur district', meticulously recorded in his diary just after Easter 1889, offers an insight into the attractions of the forest and its history.[2]

Packing (among other things) four pairs of woollen socks, eight handkerchiefs, an extra pair of boots and spare suit of clothes, four collars, a soft hat, a pair of leggings, one flannel undershirt, a pair of drawers, a guernsey and, of course, his diary, he and his friend went to Melbourne's Princes Bridge Station where they were a little concerned at the number of travellers with guns. Shooting was a popular sport. They took the train to Healesville, a pretty town with hop farms, an occasional team of bullocks (which reminded them of the gold days), and a river – the Watts – 'so transparent that you could almost see the bottom in any part of it'.

After a night in a hotel, they forwarded their portmanteaus by coach to Marysville and set off on their walking tour along the Yarra Track. 'The immense gum trees that we had heard so much about we see here for the first time', he noted in his diary, perhaps recalling the search for the giant trees that was a feature of the Centennial International Exhibition the year before. About halfway between Healesville and Fernshaw the two men noticed a grand building on the brow of a hill: '[W]e are informed that it is the Gracedale Coffee Palace and will be opened about next

Christmas'. The Gracedale later became known as 'the best furnished holiday house in Victoria'.[3]

They took delight in 'the most beautiful umbrella ferns, very much resembling palms', but also noticed that '[n]umerous fallen trees are to be seen everywhere and thousand upon thousand gaunt spectres without a leaf on them slowly decaying'. They could hear 'the low thud of the splitters' axe'. That night at Boyle's Hotel at Fernshaw, they listened to tales of the gold rush to Woods Point, and of the government reward given to the pioneers of the first narrow saddle track to the diggings. Fernshaw had quickly developed a reputation as 'a very paradise for invalids, and afford[ing] a shelter and escape from the heat and dust of the summer'.[4] But the town, they learnt, would be no more in a year's time, for it fell within the new water catchment area for Melbourne, and the government had bought out the inhabitants. Boyle, their proprietor, boasted that he received £200 an acre for land which had cost £4 an acre in 1863.

After walking to Marysville the next day, the travellers put up at a hotel that, on Easter Monday, lodged 80 people, although it could comfortably sleep only 30. It was the Australian Hotel, once a staging point for the gold escort from Woods Point and

Boyle's Hotel, Fernshaw. (La Trobe Collection, State Library of Victoria)

built in 1863 by Maurice John Keppel, who later grazed cattle on Mt Margaret and Lake Mountain. Cattlemen played a prominent role in Victoria's early mountain tourism.

The two holiday-makers spent several days exploring the forests and mountain peaks of the Marysville district, and enjoying the company of fellow guests (Federation was one of the conversation topics). They did not, however, mix with the 'silver kings', directors of the Broken Hill Proprietary Company who, they noticed, were in 'a Cook's tourist party'.

They followed splitters' tracks, travelled along a corduroy road (the continuing Yarra Track, made with earth and fallen logs laid across the track), gathered ferns, especially the uncommon ones, threw stones at a possum, inscribed their names on trees at the top of peaks, rolled giant boulders down the mountainside and were puzzled they saw few animals. There were the pleasures of history too – a 'rustic bridge', a neglected cemetery, a deserted hotel along the Yarra Track, the Cathedral Range (which evoked 'a ruined abbey') and the Australian Hotel's treasured piece of heritage: 'the identical china service used by John Batman, the discoverer of Melb.'. The travellers noted that 'it is only used on state occasions, and must be worth a great deal of money on a/c of its historical value.'

Victoria's first tourist-information office had been established at Flinders Street station just a year before this excursion into the ash forests, as a service to visitors to Melbourne's Centennial International Exhibition. The formation of clubs such as the Field Naturalists' Club of Victoria in 1880, the Melbourne Amateur Walking and Touring Club in 1894 and the Walhalla Mountaineering Association in 1907 signalled a new interest in active, outdoor recreation. At Christmas 1888 John Monash walked from Healesville to Marysville and then to Warburton and found it an 'effective brushing away of the accumulated cobwebs of a twelve-month'. Another walk took Monash along the route from Toongabbie in Gippsland through the mountains to Healesville. He went through Walhalla, which he considered the most beautiful place he had ever seen.[5]

The popular discovery of the mountains near Melbourne had begun decades earlier. In 1858, Alfred Howitt stood in wonder under tree ferns in the Dandenongs and reflected that it was 'one of those tropical looking spots one would rather expect to find in the south seas than in Australia … We ought to have been a thousand miles away from Melbourne instead of twenty, so wild and solitary was the scene.'[6]

In his history of Australian landscape painting, *Images in Opposition*, Tim Bonyhady has described the popularity of fern gullies among Australian artists in the mid-nineteenth century, and in his book *The Colonial Earth*, he has further analysed the 'Fern Fever' that gripped colonists.[7] The Dandenongs, especially, came under pressure, writes Bonyhady, 'not only from timber-getters but also from professional fern collectors and excursionists with an appetite for fern-gathering ... The colonists' fernmania became the gullies' curse'.

In the 1850s and 1860s the craze for fern collecting gathered momentum in both Britain and Australia, and paintings such as Eugene von Guérard's *Ferntree Gully in the Dandenong Ranges* (1857) received high praise and public attention. The work of notable photographers J W Lindt and Nicholas Caire also publicised the fern gullies and giant trees of the forest, and pictured an idealised life of the rural dweller. Many of their images hung in railway carriages, and a ready market existed for their work. Between 1882 and 1892 Lindt sold 25 000 prints of the Blacks' Spur area alone. In 1894 he built his home, The Hermitage, near the Blacks' Spur and

The Victorian fernery created at the Centennial Exhibition in Melbourne's Exhibition Building in 1888, an ancestor of Melbourne Museum's Forest Gallery of today (see pp. 180–1). (National Library of Australia)

received visitors at his Swiss-style chalet amidst the forest. In 1904 Lindt and Caire together wrote a *Companion Guide to Healesville, Blacks' Spur, Narbethong and Marysville*. In the forest, they considered themselves 'in the precincts of fairyland.'[8]

In 1912 the Victorian Railways published *Picturesque Victoria*, a booklet advertising tours and walks from the ends of its lines.[9] One of the most popular walking trails in the first decades of the twentieth century followed part of the Yarra Track. In 1906 the Public Works Department laid out a tourist track from north of Warburton (beginning at McVeigh's Hotel) to Walhalla, which opened access to the spectacular Yarra Falls and the Baw Baw plateau. Distance was marked on trees, and

huts were provided. Three shelters of split palings and corrugated iron – at the Yarra Falls, Talbot Peak and Mt Whitelaw – were erected along the four-day trip. By Easter 1914, a 'full house' was reported at each of them. The track enjoyed a further burst of popularity during the 'hiking craze' of the 1930s depression, but the huts were destroyed in the 1939 fires.[10]

The popularity of 'hiking' – a new word – was partly due to the writings of Victorian public servant and educationist R H Croll, even though he disdained the 'ugly verb'. In 1928 he published *The Open Road in Victoria*, a description and celebration of country walks, and was 'genuinely astonished at the success of the book'. Within three weeks a second printing was required, and Croll produced another walking guidebook, *Along the Track*, in 1930. He had joined the Melbourne Amateur Walking and Touring Club in 1897 and claimed to have walked some 3000 miles (4800 kilometres) in Victoria, many of them in the ash forests.[11] 'Mystery hikes' to undisclosed

Cover of a 1933 map showing the popular walking path from Warburton to Walhalla.

destinations from the end of railway lines became a popular, mass recreation and a new magazine, *The Hiker*, made a brief appearance in 1932.[12]

The decades following the Great War were a period of widespread celebration and rediscovery of the Australian countryside. The experience of war caused many people to recoil from urban industrialism, and it was after 1920 that the Heidelberg school of Australian impressionist painters became most revered. Tramping with book and pipe, rejoicing in the open air, seeking out 'the oldest inhabitant' of settlements, pottering around country cemeteries: these became the common holiday pastimes of literary men.

The 'hiking craze' of the early 1930s disturbed many of the established 'bush-walkers' because it seemed commercial, popular, American and feminine. 'Armies' of newcomers 'arrayed in the uniform of the road' were making incursions into the forests and mountains and it was 'the girls' who were 'leading the march'.[13] Bush-walkers considered that hikers wore sandshoes or high heels rather than solid boots, dresses and trousers rather than shorts, walked in huge, noisy groups rather than in small, skilled parties, littered and left fires alight, stuck to tracks and roads yet did not know where they were. 'It is hikers who go out and get lost; it is bushwalkers who rescue them', declared a contributor to the *Sydney Bush Walker* in 1937.[14]

The development of tourism and walking tracks often proceeded hand in hand with other uses of the forests. Miners' and splitters' tracks provided routes for walkers, and some of the timber tramways doubled as tourist trips or attractive picnic strolls almost from their inception. Timber tramways were, as we have seen, the linear habitats along which naturalists collected. Some of Croll's suggested walks utilised tramways, and one route – Yarra Junction to Noojee – took people through 'the Bump tunnel', which was the underground section of the Victorian Hardwood Company line east of Powelltown.[15] A 1913 tourist guide to the Narbethong and Marysville districts featured a cover photo of 'A Bush Fire on Mt Dom Dom'.[16] Timber towns looked to tourism as a way of surviving. Powelltown tried to sell itself to tourists as early as 1917, without success. Residents feared that 'cutting out of the forest will mean the cutting out of the town.'[17]

During the inter-war period in Victoria the bush as a place of work and as a place of recreation overlapped most clearly. Sawmillers could offer the hikers a cup of tea, or the scouts some training in bush skills. Bill Russell, the big sawmiller of the Gembrook district, had a close relationship with the Boy Scouts Association, and

Mature mountain ash forest in the Wallaby Creek catchment area.
(Photo: Tom Griffiths)

The Superb Lyrebird (see 'Lyrebird' and 'The Beauty and Strength of the Deep Mountain Valleys', pp. 46–7 and 126–7).
(Photo: Gary Lewis)

'One Mile Bridge, Blacks' Spur', photographed by Nicholas Caire about 1878–9. The huge tree on the right is 'Uncle Sam'. (La Trobe Collection, State Library of Victoria)

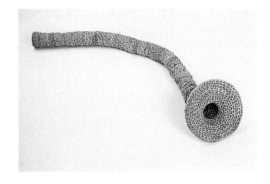

This slender eel trap crafted from sedge by Joyce Moate, a Daung wurrung Elder, is an example of those used in streams and creeks draining tall timber forests (see 'Coranderrk Calendar' and 'Shortfin Eel', pp. 58–9 and 100–1). (Museum Victoria)

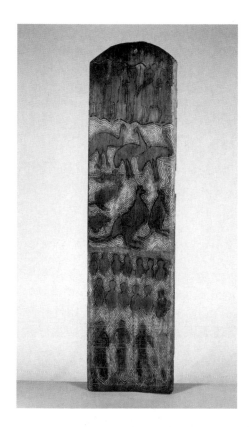

Carved grave marker for Thomas Bungaleen, c. 1865. This memorial was made by 'a man of the Yarra tribe', possibly Simon Wonga at Coranderrk, as a record of events associated with the death in 1865 of a young Gunai man, Thomas Bungaleen. (X6249, Museum Victoria)

The Morning Star waterwheel at Donnellys Creek in 1989. (Photo: Ray Supple)

Waterworn sapphires (blue to green) and rubies (to 1 cm) from gravels in Cardinia Creek (see 'Gems in the Forest', pp. 74–5). (Museum Victoria)

Waterworn topaz crystals (to 1.5 cm) from the Bunyip River district. (Museum Victoria)

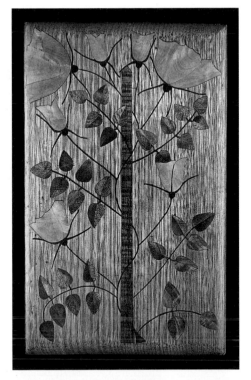

A section of parquetry using timber from E. regnans *(see 'The Good Oil', pp. 88–9).* (Museum Victoria)

A splitter's hut ('Howard's Hut') on Morley's Track near Fernshaw. Note the long crosscut saw and axes leaning against the hut. The photograph was taken by Nicholas Caire about 1879. (La Trobe Collection, State Library of Victoria)

Typical insect fauna from an undisturbed mountain stream in the ash forests. (Photo: Richard Marchant)

The Cascades at Wallaby Creek, a granite-lined channel built in 1883 and still part of Melbourne's water-supply system. (Photo: Tom Griffiths)

The larvae of the Silky Hairstreak butterfly, Pseudalmenus chlorinda, *are attended by strong smelling ants that raise their abdomens when disturbed (see 'Hairstreak Butterfly', pp. 112–13).* (Photo: Ross Field)

Museum Victoria has been researching tall forest fauna for 150 years (see 'Exhibiting the tall forests', pp. 180–1). (Photo: Petrina Peile)

Eugene von Guérard, Ferntree Gully in the Dandenong Ranges, *1857, oil on canvas, 92.0 x 138.0 cm, Gift of Joseph Brown* AO OBE*, National Gallery of Australia, Canberra*

A Field Naturalists' Club of Victoria excursion to Mount Baw Baw in 1892. (D Best's album, *FNCV Archives, copy in Historic Places*, DNRE)

Pl. 91

ZOOLOGY OF VICTORIA
(Mammals)

Coloured lithograph of Leadbeater's Possum (by Dr Wild) from Sir Frederick McCoy's Prodromus of the Zoology of Victoria, *Melbourne, 1885 (see 'Leadbeater's Possum', pp. 150–1).*

1939 was a year of fire and war. In this bushfire poster the 'enemy' looks suggestively Asian. (DNRE Library)

Didymuria violescens, *one of the species of stick insects that chews on the leaves of mountain ash (see 'Insect Outbreaks', pp. 168–9).* (Photo: Denis Crawford)

ISSUED BY THE
FORESTS COMMISSION
OF VICTORIA.

'Playing trains' on a surviving section of wooden tramway in the Murrindindi Scenic Reserve between Toolangi and Yea. (Photo: Tom Griffiths)

Timber seasoning kilns being reclaimed by the bush in the Britannia Creek valley near Warburton in 1989. (Photo: Tom Griffiths)

The Noojee trestle bridge (also known as the Cobweb Ladder) was rebuilt after the 1939 fires.
(Photo: Chris Smith)

Two quartz roasting kilns survive from early mining days in the forest. They were constructed of
rubble masonry at the New Chum mine near Dry Creek, probably in the late 1860s.
(Photo: Ray Supple)

Baden Powell was once a visitor at his Swallowfield property.[18] When a party of naturalists sheltered from a hailstorm in a sawmill, the workers showed off their handiwork of whittling walking sticks from blackwood.[19] Walkers were tourists in time as well as place, and they were pleased to follow historic routes and to find old bush inhabitants who could tell them of the past. The urban escapees romanticised the male bush worker and generally welcomed his presence in the forests. There had not yet developed that enmity that we know today between bushwalking conservationists and timber workers.

Perhaps the most famous member of the Melbourne Walking Club (and also a secretary and president) was Bill Waters. His first walk, as it was for many, was along the Warburton–Walhalla Track in 1917, with a chaff bag slung on his back. Two years later he walked with a friend from Toongabbie 'along the old corduroy coach road, formerly Campbell's Track, to Walhalla, and thence to Aberfeldy, Red Jacket, Jericho, Matlock, Woods Point, Jamieson, and so on to Mansfield.' The following year he again visited some of these 'ghost towns' as well as Toombon, Donnellys Creek, Edwards Hill and the Springs. He camped by the ruins of deserted hotels and enjoyed meeting some of the old timers and got them talking about the old days. He 'picked up a few relics of old Matlock, viz., some burnt copper nails' and on another trip in the 1920s salvaged three partly burnt books recording some of the early history of Donnellys Creek. There were still gold fossickers about and when one old digger saw the walking club party arrive he asked: 'What's this? Another flamin' gold rush?'[20]

Tourists were not yet easily distinguishable from bush workers. Some were attracted to the romantic notion of the swagman as a free Australian spirit wandering the bush tracks and were delighted when they were mistaken for one. 'Want a job?' some farmers asked.[21] Later walkers developed specialised equipment that made them distinct, but the first mountain hikers were without even good maps and sometimes had nothing more precise than *Broadbent's Road Map of Eastern Victoria*.[22]

The Melbourne Women's Walking Club was formed in 1922 (women could not be members of the Melbourne Walking Club), and some men resented this invasion of what they saw to be their domain: the bush.[23] The club's first walks were in the ash forests. The women admired the few great trees that remained and explored the tracks that miners and splitters had made. They took an interest in the signs of industry that they passed. Some even rode on the skyline leads that carried logs

A Melbourne Women's Walking Club excursion to the Upper Yarra in the 1930s (at the junction of Woods Point and Reefton Spur tracks). (Historic Places, DNRE)

across the valley. Unevenly spaced sleepers along tramway routes or high bridges with a plank missing were bushwalking hazards. Unoccupied mill huts offered welcome accommodation and, if you were lucky, 'you might even snuggle your sleeping bag close to a still warm boiler at the mill'.

Even before the formation of the club, women were doing long walks through the forests of ash. In 1909 Ethel Luth and two friends walked 224 kilometres from Warburton to Alexandra via Matlock and Woods Point. She recalled that women on the track were 'a decided novelty ... news of our coming usually travelling ahead of us'. When they were walking through the old mining town of Gaffneys Creek, a woman waved to them from the verandah of her cottage. She had heard that the walkers were coming and had prepared afternoon tea for them.[24]

Guesthouses proliferated in the mountain valleys, particularly between the wars. Healesville, Warburton, Marysville and the Dandenongs boasted stylish accommodation for thousands. Gracedale House, noted by the Marysville excursionist, was one of the most famous, and its life (1889 to 1955) spanned the full era of the guesthouse.[25] Even the 1939 fires did not daunt the Gracedale's proprietors, who were advertising even as the flames circled the town. Many of the guest-houses, both in the mountains and by the sea, drew customers seeking health as well as a holiday. A cheaper form of accommodation, Australia's first youth hostel, was established at Warrandyte at the beginning of the Second World War.[26]

Although Healesville was a popular outing for many of the new car clubs established between the wars, cars eventually opened up wider and more distant tourist horizons for the city traveller – and for the bush resident. Alan Mickle remembered Narbethong near the Blacks' Spur in the early twentieth century: '[M]any of the dis-

A walker rides a logging skyline. (Historic Places, DNRE)

trict's inhabitants had never seen a train', he recalled. 'More had never seen the sea.' He remembered when the first motor vehicle came over the Blacks' Spur. 'It was a high-wheeled motor buggy painted a brilliant red. It was terrifyingly noisy … it shattered the stillness of the bush for miles around. Every dog in the district woke up and began to bark furiously … A bush fire raging over the country-side could not have caused very much more commotion.'[27]

Many guesthouses were mysteriously burnt during the depression of the 1930s when expensive holidaying was curtailed. From the 1930s, as the holiday fashion became more lively and informal, Healesville was warned that it could no longer rely on its 'teddy bears and scenery'. This was partly a reference to the Sir Colin MacKenzie Sanctuary for Australian fauna, which had been established with the support of the Australian Institute of Anatomical Research in the early 1920s (and officially opened in 1934) on part of the site of the Coranderrk Aboriginal Station.[28] There were sinister continuities between these two reserves for indigenous life, both very popular tourist sites in different eras, one a 'show place' of Aboriginality and 'primitive' humanity, and the other a sanctuary for 'low-degree, slow breeding and altogether archaic mammals'.[29] Both enclosed 'remarkable survivals' and con-stituted Australian research parks in 'the problem of world evolution'. One of the platypuses reared in captivity at Healesville was called Corrie, an abbreviation of Coranderrk.

After the Second World War, people increasingly turned to the beach for holiday informality, partly because of changing perceptions of the sun and its benefits, and partly too because it was just at this time that the mountains began to expand their winter recreational facilities.[30] Although skiing was a popular pastime for miners at Kiandra in New South Wales from as early as the 1860s, it was not until the turn of the century that it was actively encouraged in the Victorian mountains, and only between the wars that resorts developed.

From 1906 people in Walhalla advocated Mount Erica as a rival to other skiing areas such as Mount Buffalo, where snow sports were encouraged in the 1890s. In the early 1900s Mount Donna Buang near Warburton was 'Melbourne's great unknown mountain', 'a mountain that is not shown on any map of Victoria', but by 1912 a graded bridle track was opened to the summit and soon skiers were using it to reach the mountain's snowfields.[31] Locals made their own skis by selecting planks of a suitable size from one of the many timber stacks, smoothing them on a lathe and then steaming them over a boiling copper to give their ends a curve. They then skied the mountain's fire-breaks with these 'home made sticks of mountain ash' strapped to their boots.

By 1933 an additional 500 cars sometimes were making their way from Mel-bourne to Mount Donna Buang on a Sunday.[32] On the Baw Baw plateau, ski runs were cut by a local Rover Scout crew in the mid-1930s, and in 1944 the Mount Erica division of the Ski Club of Victoria was formed. The 1939 fire opened up the slopes nicely for them.[33] In 1963 the Country Roads Board completed a new road to the Baw Baw Alpine Village site, where lodges were built and downhill runs and tows developed. In the late 1960s a proposed tourist road across the Baw Baw plateau was halted by strong community opposition and concern for conservation. With the surge of interest in cross-country skiing since the 1970s, Lake Mountain has become another popular snow resort in the forest. Sometimes 10 000 visitors ply its slopes on a winter weekend.[34]

The declaration of national parks, which was frequent in the first and third decades of the twentieth century and again in more recent years, was a measure of a growing enthusiasm for nature. However, it was the demands of outdoor recreation as much as any conservation ethic that shaped these early reserves.[35] Often these roles were in tension, as is suggested by the term 'park'. Part of the Fern Tree Gully National Park was first reserved in 1882 as a site for public recreation; it quickly developed as a focus for city outings. In the subsequent 100 years, many visions and changes have been imposed on it, few of them with the preservation of native flora and fauna as their end. In 1936, for example, a zoo with native and exotic animals opened in the park, and between 1932 and 1940 over 300 ornamental seedlings were planted. Proposals to build chairlifts, golf courses and an open-air theatre posed the dilemma of the purpose of the park more sharply. Every nature park is thus a museum of human perceptions.[36]

Until the formation of a National Parks Authority in 1956, parks in Victoria were administered by individual committees of management, composed almost entirely of unpaid citizens with little access to finance or trained staff. Such groups necessarily focused on the fund-raising and recreational aspects of management and often faced the dilemma of allowing the continuation of exploitative activities within the park or denying themselves a much needed source of revenue. In Kinglake National Park, created in 1928, the committee of management had to permit the felling of timber for firewood in order to fund protection of the rest.

Since the 1960s, the growing dominance of ecological principles in landscape evaluation introduced a biocentric rather than anthropocentric focus to park management. The first conservationists had been champions of the 'wise use' of resources; then there developed a movement to preserve aesthetically pleasing or spiritually uplifting places. The ecological vision shifted the emphasis to non-human values: the preservation of biodiversity, the protection of gene pools, the integrity of ecosystems, the independent rights of animals and plants. These evolving priorities were reflected in the sorts of national parks that Australians set aside from late in the nineteenth century: at first they were 'wastelands' or areas of little perceived economic value; then tracts of outstanding scenery or places of urban recreation; and, most recently, areas of biological richness or rarity, regardless of their scenic or recreational appeal.

Nature conservation came to have purposes completely independent of tourism and sometimes in conflict with it. The rise of a wilderness aesthetic also cast humans as intruders, and projected an unacknowledged and problematic historical vision of

'THE BEAUTY AND STRENGTH OF THE DEEP MOUNTAIN VALLEYS'

ELIZABETH WILLIS

Melbourne's first European settlers turned their backs on the forests east of the town, finding the wide, open spaces of the inland plains more tractable and productive. The stands of mountain ash at first seemed inaccessible, dense, dark and alien. Gradually, however, Victorians discovered the beauty of the tall trees, and began to collect things and develop symbols that evoked the mountain ash environment. In this way the forest began to encroach upon the city.

The first experience of the forest for many Victorians was in an annexe to the Exhibition Building at the 1880 International Exhibition, where a fernery 'designed to reflect a Victorian fern gully' promoted tourism to Fern Tree Gully and Fernshaw. Soon afterwards, day visitors were able to catch a train to the Dandenongs, explore the forest, and dig up ferns and wildflowers for display in their rooms and gardens back home.

Motifs of lyrebirds and ferns became popular as interior decoration. The people of Coranderrk took advantage of this fashion, hunting lyrebirds and selling the tail feathers to Europeans to adorn their suburban drawing rooms. Some visitors collected ferns and grasses, and pressed them into scrapbooks. This was the great age of landscape photography, and stereographs brought the immediacy of three-dimensional views of waterfalls and forest walks into Melbourne's homes. The forest images created by photographers like Nicholas Caire and artists like Eugene von Guérard shaped Victorians' view of the landscape.

Lyrebirds, ferns, gum leaves and nuts joined the kangaroo and emu as distinctively 'Australian' icons, appropriated as part of the growing nationalism. Ferns appeared in the iron lacework which decorated the verandahs and balconies of both cottages and mansions. They were etched on souvenir mugs, tooled in leather on bookbinding, and moulded by silversmiths to form mounts for countless presentation emu eggs.

Lyrebirds were popularised on stamps and on chairs, and they were a prominent motif on the arch erected by Victoria's German community to commemorate the opening of the first federal Parliament in Melbourne in May 1901. Lyrebird feathers even made a brief appearance as an element in an unofficial Australian coat of arms. Ferns, lyrebirds, wattle and heath were extensively used by graphic designers on illuminated addresses, certificates and medallions. Designers, artists and architects deliberately used images of Australian flora and fauna to promote national feeling and to develop a 'national style'.

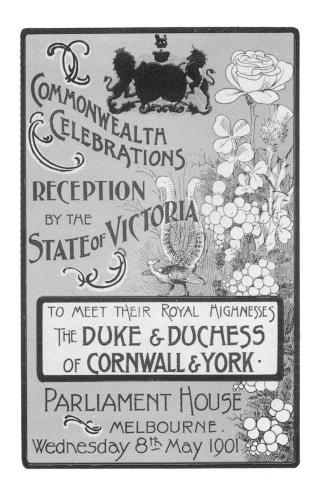

A Lyrebird shown with wattle, and the British symbols of rose, shamrock and thistle, to commemorate the Royal Visit in 1901. (Royal Historical Society of Victoria)

The forest environment seemed to be particularly pure and uncontaminated. The imagery of fern gullies and lyrebirds was especially popular with temperance organisations, which promoted purity of life and avoidance of alcohol, and with organisations aimed at Australian children. Words reinforced the imagery of peace and purity. Henry Kendall's poem 'Bellbirds', which was very popular in the early twentieth century, was originally written about the northern rainforests. However, Victorians transferred his poem to the mountain ash landscape, and identified with his description of the forest as a source of refreshment and rejuvenation.

Images from the forest are still powerful. The logos of the National Trust, the Shire of Sherbrooke and the Wilderness Society all employ forest symbols: gum leaves, lyrebirds, a platypus. Such symbols still evoke feelings of pride in Australia's forest landscape, and generate support for its preservation.

landscape. Many of the definitions of wilderness that emerged in the 1980s were sensory; they concerned boundaries of sight, lengths of walks in one direction, and the maintenance of an authentic sense of human danger and isolation. They measured remoteness and absence in human terms. A human presence in the past or the present detracted from perceived natural values and this history was therefore often denied by the politics of conservation. Yet, as ecologist Ian Lunt has argued, a historical perspective is essential in managing the conservation of dynamic ecosystems. Furthermore, the 'naturalness' of past landscapes is no guarantee of biodiversity. 'Biodiversity conservation', he concludes, 'must involve human interaction'.[37] Those 'intruding' humans are beginning to find a natural role for themselves in the forests again.

10

BLACK FRIDAY

Great fires become secured in folklore, grimly named by their day of terror. 'Black Thursday' is the name given to the 1851 fire, 6 February, when it seemed that the whole of Victoria was alight and ash fell on ships at sea. 'Red Tuesday' recalls 1898, 1 February, when fires raged in Gippsland and more than 1500 people lost their homes. 'Black Sunday' designates the 1926 fire, 14 February, when 31 people died at Warburton in a single day. 'Ash Wednesday', a name ready-made, commemorates the sacrifice and tragedy of 16 February 1983, when fires in Victoria and South Australia claimed 71 lives and burnt 2528 homes. However, it was the fire of 1939 that burnt itself most fiercely into memory: 'Black Friday', Friday 13 January. Named by a day, it represented a summer of destruction.[1]

It was a long, hot summer and it followed a dry winter. During 1938 rainfall across Victoria was two-thirds of the normal total, and rivers and creeks were at their lowest levels in 80 years. In the forests the eucalypts dumped their leaves and bark in massive quantities. The undergrowth was dry and the litter 'as brittle as cheese'.[2] The ground crunched ominously beneath the boots of bush workers. Temperatures reached 100°F (38°C). Early in January a high pressure system settled on the Queensland coast and a low pressure system arrived off north-western Australia. The ridge of high pressure that built up between them drove hot northerlies from Australia's arid centre across Victoria. Hundreds of little fires that had been smouldering and burning for weeks were given new and furious life. They surged together into a giant conflagration.[3]

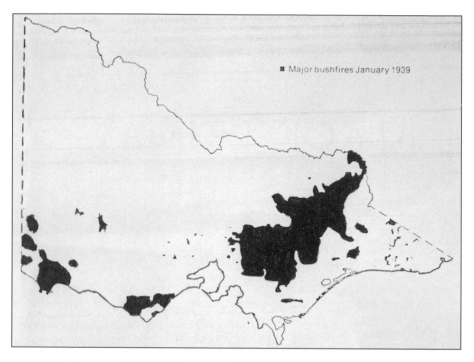

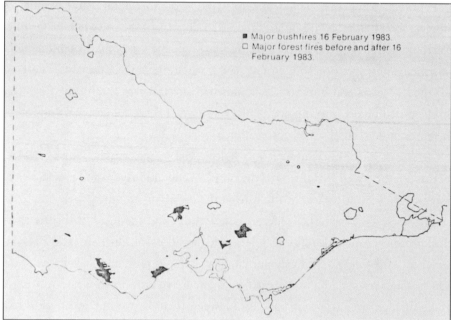

The two maps show the extent of devastation caused by Victoria's two major twentieth-century fires: Black Friday 1939 and Ash Wednesday 1983. (Reproduced from the Victorian Yearbook *1984, Government Printer, Melbourne, 1984.)*

Near the township of Kinglake, north of Melbourne, a small fire had begun with the new year, probably on private property. By 2 or 3 January it was still unattended and entered state forest. On Sunday 8 January, Toolangi district foresters Charles Demby and John Barling were leading a team fighting the fire and 'burning back' when Barling, the senior forest officer, went to get a fresh view of the threatening flame. The wind suddenly strengthened and changed, and he was isolated and caught. Demby ran to warn him, and both were cut off from the rest of the party and engulfed. The following morning their bodies were found among the ash. Demby's son Alex, then aged 16 and one of the firefighting party, recalled that 'when we found their bodies later, my dad had been carrying Mr Barling'.[4] Barling's watch had stopped at 1.20 pm. They were the first victims.

Their funerals, like their lives, became hostage to the fires, which rapidly intensified. As the bells of Healesville solemnly tolled their respect, fire bells pealed urgently in bush towns along the Great Divide. Panicking and frightened farmers added to the flame by burning-off on their own land to protect themselves from uncontrolled wildfire.[5] Charles Demby's funeral was delayed an hour as his family negotiated a way around another fire on the Toolangi road. In the Rubicon range to the north about 110 men, some of them with families, were at work in the bush sawmills. There might have been more, except that some were still on their Christmas holiday. Twelve people died as the fire rushed down upon them.[6]

On Tuesday in Melbourne it was the hottest day since 1862. The temperature reached 112°F (44°C). On Black Friday humidity reached a record low of 8 per cent. The city did not miss the chance to profit from the bush:

> Flights over the bush fire areas were being offered by aircraft operators at
> Essendon aerodrome yesterday. Visitors were met by ticket sellers with the
> invitation, 'Take a flight over the city or over the bush fires, sir?' Aeroplane
> trips to the burning forests were quoted at 30/–.[7]

Sightseers and souvenir-hunters from Melbourne began to loot burnt homes along the Yarra.[8] Strong northerlies intensified the fires and were intensified by them. Some spoke of the fires roaring like an express train. Timber-cutters working out of Woods Point told how they could not stand upright in the face of the gale. At Warrandyte, people fled to the Yarra River while three churches burnt and 100 houses were destroyed. At Woods Point old residents initially ignored the huge pall of smoke coming over the north-eastern hill; they had seen it often enough. Within an hour their town was incinerated.[9]

In Noojee the postmistress, Mrs Sanderson, sent out messages from the wooden post office as the fire burnt through the town towards her. She locked money and valuables in the post office safe and had her brother wire the keys to her wrist. She sent out the final message as flames licked the building, and then rushed to join 60 people immersed in a pool of the river behind the hotel, where they stayed for five hours.[10] At Nayook West, near Powelltown, nearly 30 people sheltered in 'the Bump tunnel', an underground timber tramline. Others retreated to mining adits and survived. At Tanjil Bren, Ben and Dorothy Saxton and Mick Gory died in a small sawmill dugout. At Fitzpatrick's mill in the Matlock forest virtually the whole of a small bush community – 15 people – died.

In a eucalyptus forest on a hot summer day, with the bush tinder dry and a vicious northerly whipping the tree tops, there was only one way to go – down. Dugouts should have been provided at sawmills. These were trenches in the ground or in the

Battling the Black Friday fire at Erica. (DNRE Library)

side of embankments, generally supported with corrugated iron sheeting and timber props heaped over with earth. They had one narrow opening, which could be shielded with a blanket that was to be kept constantly wet from the inside during a fire. Water, food and first aid equipment were to be stored inside. All too often, however, no dugouts were provided for workers and their families at sawmills. The dugout that saved the lives of workers in the Ada forest had been built only over the opposition of the manager, who believed that fire would never invade the cool, moist southern side of the range.[11] In spite of the tragedies and loss of life in the 1926 and 1932 fires, there were still many sawmills without dugouts. They were the sawmiller's responsibility. They were to be built and stocked at the sawmiller's expense. Many didn't bother.

So when the fire bore down upon them, timber workers buried their belongings and tried to bury themselves. Some made makeshift dug-outs, which became death traps. Some burrowed into the sawdust heaps and made themselves an awful, suffocating tomb. Some jumped into water tanks and were boiled. The few who did survive without well-equipped dugouts were near big, broad creeks in which they could immerse themselves, or else they were able to find a large cleared area, in the centre of which they lay down, wrapped themselves in wet blankets and kept their nerve. The trauma of Black Friday drove some people off the edge of sanity for the rest of their lives. When the survivors tried to walk out to civilisation after the fire had passed, they found themselves lost. It was a different landscape. All familiar things had gone. The matchstick trees were no guide to where they were.

It took courage and desperation to climb underground 'like a wombat' and stay there. Even where dugouts existed, many bushworkers did not trust them. At the Ruoak No. 3 mill in the Rubicon forest, workers crammed the dugout with their furniture and fled. The four slowest died. 'Some of their furniture is still there today', reports historian Peter Evans, 'embedded in the collapsed remains of the dugout'.[12] Ruby Lorkin survived the fire in a dugout at the Ada No. 2 mill near Powelltown. She recalled:

How we lived through that dreadful inferno I shall never know ... We two women were told to lie down on the ground whilst water was thrown over us, until the water reached boiling point and we could not drink it. Four men at a time stood at the small opening of the dug-out holding up soaking

blankets until the blankets dried, caught alight and were swept from their
hands in a few moments, then another four went forward to take their place.
The Engineer went berserk and tried to take his wife outside and had to be
quietened by knocking him unconscious to save his life.[13]

The fire engulfed the forests and farms from the Murray River in the north to the
Princes Highway in the south, sweeping across the wedge of mainly mountainous
land in eastern Victoria. Fire also broke out in the west of the state, in the Otway
Ranges, in the Grampians, along the south-western coast around Warrnambool and
Portland, and in South Australia and the Australian Capital Territory. The arithmetic
of disaster, impressive as it was, hardly captured the enormity of the experience: 1500
people were left sheltering in temporary camps and homes; 69 sawmills were destroyed and another 14 damaged, most of them in the ash-milling belt where the
greatest volume of sawn timber was then being produced; 126 kilometres of bush
tramways were lost; 55 bridges, 80 horses and 700 houses were burnt, as well as a
hospital and ten guesthouses and hotels; Narbethong, Noojee, Woods Point, Nayook
West and Hill End were completely gutted, and Warrandyte and Yarra Glen partially
destroyed; 1.4 million hectares were burnt; and 2070 million super feet (nearly 5
million cubic metres) of timber were destroyed in the central highlands alone, representing 20 years of potential sawmilling in that area.

   Another catalogue of disaster can be compiled: one of stories and images, visual
and sensual experiences, the words people used as they stumbled out of the blackened
forests, the hooks upon which they later hung their memories. There was the strange
behaviour of matches, which served as both warning signals and telltales, for they
burnt white for days before the great fire, so they said, and blue afterward in an
atmosphere charged with carbon dioxide. There was the thick pall of smoke that
turned day into night, the ash that fell at sea, in Tasmania and in New Zealand. There
were the deafening roar and blasting windstorms of a freak forest fire, the tornadoes
that ripped trees off at the ground, the explosions of gas, the fire leaping kilometres
ahead. There was the vulnerable innocence of those who did not know the scale
of what they were fighting until it was too late. There was the machinery at bush
sawmills that became a molten mess. There were the dugouts that became tombs for
some and saviours for others. People told stories of taking it in turns to hold wet
blankets across the dugout doors until the skin on their hands and faces curled back.

Suspended narrow gauge rails and smouldering stumps were all that remained of this trestle bridge, one of several burnt near Noojee in 1939. (Argus, 16 January 1939)

There was the dead silence of the day after, with not a bird or animal or leaf to stir, and the creeks running black as ink. And there was the mist that seemed for years afterwards to hang low over the forests of ash, a mist made up of bleached, dead spars, the skeletons of the forest.[14] The Red Cross, 'concerned about the health of the bush fire refugees' as they emerged from the smoking forests, appealed to the public for 'gifts of tobacco'.[15]

Black Friday was not a freak event – it was one of those catastrophes endemic to the ash forests – but it had distinctive European dimensions. It was a cultural creation, a culmination of a century of white settlement and environmental practice. There had been warnings, which had been gathering apace in the inter-war period. The fires of 1926 and 1932 had been severe ones, and there had been those of 1898, 1905, 1908, 1914 and 1919 before. In the forests of ash it was the frequency – and not so much the intensity – of fires that was a result of European settlement.

Black Friday was also a European creation in a more immediate sense: *'These fires were lit by the hand of man.'* These were the words of Judge Leonard Stretton, who conducted the Royal Commission into the causes of the 1939 fires that was set up within two weeks of the disaster.[16] A film made by the Forests Commission soon after Black Friday was titled *The Hand of Man,* and the camera zoomed in on Stretton's single sentence indictment, underlined.[17] It was society and not nature that was under trial. Stretton highlighted 'the indifference with which forest fires, as a menace to the interests of all, have been regarded.'[18] Fire was someone else's responsibility. It was, as one witness to the commission put it, 'nobody's business to put out'.[19]

And who were the firebugs? Rarely were they malevolent arsonists. Mostly they were farmers and bush workers, and their fire-lighting was sometimes casual and selfish, sometimes systematic and sensible, and increasingly clandestine and rebellious. They were ordinary people going about their lives who had not learnt the potential of fire and were careless with it, or who feared wildfire and wanted to pre-empt it. They were settlers burning to clear land, graziers firing the grass to promote new growth, miners blazing a path to a new reef, jackeroos signalling their where-abouts to their bosses. Burning was a rite – and a right. They were home owners who, when they saw smoke on the horizon, threw a match over the back fence. A newly burnt home paddock was like a safety blanket, a protective measure.

In 1855 William Howitt had described the Victorian gold-rush populace as 'this fire-scattering race of rude men'. The diggers, he said, 'burn up the country wherever they go, as they say, to get rid of snakes.'[20] Almost a century later, sawmiller Jack Ezard used the same language: 'There is always somebody foolish enough to light fires. I have seen people burn snakes.'[21] But he reminded the commission that 'These people have to burn the scrub to live.' Although Ezard, born and reared in Gippsland, acknowledged that lighting a fire at the wrong time could be a criminal act, he also insisted that 'I think it is almost as criminal an act not to light a fire at the right time'.[22] When travelling from Powelltown to Yarra Junction it was normal to see 'half a dozen fires on the sides of mountains'.[23] 'The whole Australian race', summed up one witness to the Royal Commission, 'have a weakness for burning.'[24]

The Royal Commission into the 1939 fire accepted this finding. It was Australian society and not just Australian nature that was dependent on fire. Stephen J Pyne has described how the bush dweller – like the Aborigine and like the eucalypt – lived with fire, and could make a living because of fire.[25] Perhaps fire was so much a part of the

Australian landscape and character that it could never be eliminated or suppressed. It had to be accepted and used, and perhaps it could be controlled.

Stretton was not the first to condemn Australian carelessness with fire: he joined a long tradition. In 1890 George Perrin, Victoria's first Conservator of Forests, described the 'universal carelessness with regard to fire' as 'culpable negligence', and annual reports of the Victorian Forests Department and later Forests Commission echoed these words.[26] However, the 1939 Royal Commission gave official recognition to a folk reality and tried to give direction and discipline to a widespread popular practice: burning to keep the forest safe. It recommended that the best protection against fire was regular light burning of undergrowth at times other than summer. Only fire could beat fire.

Foresters aligned themselves uneasily with popular practice. They saw themselves as 'experts' and educators, and many of them were born and bred in the city. Farmers and bush workers were often suspicious of their bookish knowledge. In the words of C J Dennis, they distrusted 'the Theorists' who controlled 'that bit of Australia that these elder bushmen know and love'.[27] Foresters, in turn, recognised a cultural gulf between themselves and the locals. D M Thompson, a forester in the Cann River and Mallacoota District of East Gippsland, wrote of his predicament in 1949: 'The more I learn ... the clearer I see how dependent a forester is on local knowledge ... I consider that no matter how efficient an officer may be, he needs staff skilled in local lore.' However, public relations were fraught with difficulties, particularly in remote areas. 'Cut off largely from outside contacts,' continued Thompson, 'public opinions are parochial in the extreme, new ideas being viewed with suspicion.' He considered that many of the farmers had 'an astonishingly limited knowledge of the bush around them'.[28]

Nevertheless, Thompson made it his task to plumb that local lore and later wrote a forestry thesis in which he described traditions of burning in the Cann River Valley handed down over three generations.[29] Foresters carefully distinguished 'control burning' from these traditions of 'burning off'. Burning off was the tool of graziers and farmers, an annual maximum-intensity fire generally with a destructive or productive aim. Control burning was a scientific technique of the forester, a low intensity, less regular fire for protective purposes.

At the Royal Commission in 1939, the youth and naivety of forestry officers ('young city men') came under attack from older bush people. D M Thompson, a 23-year-old forest officer based on the Blacks' Spur, was one whose age drew comment from Stretton: 'Apparently the local boys, who are born and bred in the bush, have to ask his permission before they do anything'.[30] Noojee timber-getter Peter O'Mara declared: 'I do not think any man should be put at the head, or have anything to do with the head of the Victorian State Forests unless he is a practical bushman, and has been for many years'.[31]

The Forests Commission as a whole was under attack and was portrayed by the media as 'Enemy No. 1'. Stretton rebuked the newspapers for their campaign against the Forests Commission, warning them that they could 'drive me as one who is very human and loves justice, to be very one-sided and unfair, and to come to the rescue of the Forests Commission and its officers, who are being daily mendaciously and libellously assailed by the press'.[32] But Stretton himself warned the Forests Commission (and the Board of Works) about manipulating evidence brought before him, of orchestrating 'a solemn farce' by putting forward hand-picked witnesses. Certainly the Forests Commission did prepare and coach its witnesses. Said the judge: 'I do not want you to think that yesterday, although it was hot, I was asleep, and that you could do anything without my noticing it'.[33]

The chief complaint against the Forests Commission was that its management of the forests had introduced an overly zealous fire suppression regime which discouraged and even outlawed the safe, regular burning of the forests such as had been carried out in 'the graziers' time' and 'the aboriginals' time' in the settlers' chronology. They considered this a 'nightmare' policy and they dreaded the Forests Commission's 'dirty forests'. Some forest officers, it was claimed, did not know what a wildfire might be expected to do. Landholders claimed that they were now afraid to burn even when it was safe: '[A]s soon as there was a bit of smoke Mr Demby was down on you' recalled a Yarra Glen farmer of the forest officer whose death had introduced the statewide tragedy.

In the weeks after the fires, farmers and graziers held public meetings critical of the Forests Commission's policy of 'leaving all the bark and debris about', and senior forestry officers were just as quick to accuse landholders of burning off in dangerous weather. The Board of Works was critical of the Forests Commission for not controlling fires in state forests with sufficient vigour such that they invaded adjoining water catchment areas, and forestry officers accused the Board of allowing 'dead, dangerous timber to accumulate in some catchment areas'.[34]

George Purvis, a storekeeper and grazier at Moe in Gippsland explained that the forest officers are 'so keen that they hate to see even a little gum tree destroyed and that is where we think differently from them'. Everybody used to burn off many years ago, he explained:

[W]e could meet a few of our neighbours and say "What about a fire" ...
Nowadays, if we want a fire we nick out in the dark, light it, and let it go.
We are afraid to tell even our next door neighbour because the Forests
Commission is so definitely opposed to fires anywhere, that we are afraid
to admit that we have anything to do with them.

As a result, he explained, the bulk of farmers did not burn their land. And so, as fires gathered force in the week before Black Friday, people desperately burnt to save their property and their lives, and these fires (indeed 'lit by the hand of man') 'went back into the forest where they all met in one huge fire'. 'To commit suicide on a Friday is thought by some to be unlucky', reflected Judge Stretton later. Particularly on the thirteenth of the month. Yet such were the fires that had been unleashed by Victorians on that day that it 'appeared to be a wicked attempt at State suicide'.[35]

Within weeks of Black Friday, the Commission heard evidence in Melbourne, Healesville, Kinglake, Marysville, Alexandra, Mansfield, Woods Point, Colac, Forrest, Lorne, Willow Grove, Noojee, Maffra, Omeo and Belgrave. Judge Stretton was very keen to get out of the city as early as possible and hear evidence from the bush. He toured disaster scenes amidst ash and dust, held court in temperatures of over 100°F (38°C), and continually reassured those giving evidence that he was there not to apportion personal blame but to discover general causes.

The transcript of testimony is a wonderful neglected source of environmental experience and sensibilities. Informal dialogues spring from the official page. At times, one can see the raising of an eyebrow. As the country hearings progressed, the judge declared that 'it is impossible for us to go to many more places, because if we do ... I think we should be on the road for some years, like Barnum's Circus, and be about as ridiculous'.[36]

Although Stretton was a staunch upholder of the dignity of the law (he rebuked one shire council for summoning him as if he was 'a sort of judicial taxi to be hailed at the word of command, or even whistled'), he nevertheless bent over backwards to put his witnesses at ease and to empower them ('Do not agree with what I am saying because I am saying it', he urged one 'bushman').[37]

Although Leonard Stretton (1893–1967) was, in his own words, 'born in Brunswick only a lifetime after the settlement of Melbourne', for three years from the age of six he had a rural upbringing at Campbellfield on the outskirts of the city, and it was a happy and formative period of his life. His father had been a teetotal brewery clerk, publican and compulsive gambler, and the young Stretton felt at home with working-class rural culture. He became a champion of the underdog and an advocate of bush values.

One after another the witnesses testified to the use of fire and to changes in the forest since European settlement. They discussed the forest floor, what was 'litter', what was 'scrub' and what 'rubbish', what was 'dirty' and what 'clean', and how it came to be that way. The commission pursued these definitions. Was 'rubbish' dead or alive, or could it mean scrubby green growth as well as drying debris? Did frequent fire eradicate rubbish or bring it on? And, anyway, 'what is the fire you would call a fire?' as one witness was asked. Is it anything that would throw up smoke, or a campfire or burning off or wildfire? What would denote 'carelessness' with fire – accidental ignition, thoughtlessness, arson, casual and selfish fire, systematic and wilful fire, burning at the wrong time of year, illegal burning at the right time of year? The Royal Commission was nothing less than a full-scale enquiry into Australian bush culture. The language the forest workers used – 'burning to clean up the country' – was uncannily like that of Aboriginal people.

There were two possible cures for scrubby, fire-prone forests. One was the long and total exclusion of fire, because fire induced scrub growth. The other – a shorter-term policy – was to introduce regular, controlled fire. The first, Stretton concluded, was impossible in European Australia. The second was realistic.[38] The first was a view put forcefully by Alexander Kelso on behalf of the Board of Works, which had particular reasons to manage a mature forest to maximise water harvesting, and also by C E Lane-Poole, the quintessential European imperial forester who became the first Conservator of Forests in Western Australia in 1916 and the first Common-wealth Inspector-General of Forests. Lane-Poole believed that the total exclusion of fire would enable natural succession to proceed, resulting in less undergrowth and a less flammable forest.

The idea of succession to a stable climax community was central to the work of the American prairie ecologist, Frederic Clements, whose ecological model became influential in the decades before Black Friday and directed scientific studies of plant communities over time towards the expectation of an ever-increasing stability. A disturbance like fire would block the natural succession. Stretton, the Australian jurist, repeatedly questioned Lane-Poole, the European scientist, about how long the expected transition to fire-free stability would take. The judge clearly doubted that fire could be totally excluded from an Australian forest. As Stephen Pyne has observed, ultimately the dispute between Stretton and Lane-Poole went beyond conflicting hypotheses about fire and the evolution of unburned bush, and 'involved differing perceptions about what it meant to be an Australian and about how Australians should live'. Stretton perceived that there was a social ecology of fire.

The impact of the commission's report was partly due to the literary skills of its author. Leonard Stretton was a County Court Judge and declined appointment to the Supreme Court, preferring his duties as Chairman of the Workers' Compensation Board and as a Royal Commissioner (he served in this capacity on five occasions).[39] Judge Stretton was renowned for his wit and poetry as well as his forthright pronouncements, and was sometimes called Victoria's 'judicial bard'. He was a noted raconteur and a lover of raffish literature. He treasured a full edition of Charles Dickens and had a particular liking for C J Dennis's *The Sentimental Bloke*, an epithet he might privately have given himself.

His report into the causes of the 1939 fires was acknowledged as a literary masterpiece, anthologised in a collection of Australian nature writing, and passages from it became a prescribed text in Victorian Matriculation English.[40] His use of the passive voice in the report dramatised the vulnerability and helplessness of bush workers. Yet he made it clear that what seemed inevitable was actually the unseen social and natural consequence of individual acts. Stretton's empathy for ignorance resolved into a sermon of responsibility:

Men who had lived their lives in the bush went their ways in the shadow of dread expectancy. But though they felt the imminence of danger they could not tell that it was to be far greater than they could imagine. They had not lived long enough. The experience of the past could not guide them to an understanding of what might, and did, happen. And so it was that, when millions of acres of the forest were invaded by bushfires which were almost State-wide, there happened, because of great loss of life and property, the most disastrous forest calamity the State of Victoria had known.

These fires were lit by the hand of man.

It was a powerful, moral vision and the language was biblical. There were hints of culpability, almost of evil. Of the disaster at Fitzpatrick's mill Stretton said: 'The full story of the killing of this small community is one of unpreparedness, because of apathy and ignorance and perhaps of something worse'. In Stretton's 1946 Royal Commission on Forest Grazing 'the hand of man' again makes an appearance, and again it was social life and morality that were under trial. The tendering of grazing licences, he complained, took no account of the reputation of the licencees. 'The problem of forests, soil and water', he concluded, 'are problems of the behaviour of mankind.'[41]

George Sellars, the lone survivor of the 1939 fire at Fitzpatrick's Mill in the Matlock forest. (From Bush fires: A pictorial survey of Victoria's most tragic week, January 8–15, 1939, Sun News-Pictorial, Melbourne.)

However, there were always two hands at work, 'the hand of man' and another 'giant hand', a retributive hand: 'for mile upon mile the former forest monarchs were laid in confusion, burnt, torn from the earth, and piled one upon another as matches strewn by a giant hand.'[42]

What was to happen to these matchstick trees? As well as mourning the loss of human life, Victoria mourned the destruction of vast quantities of best quality timber, trees that it would have taken 20 years to mill. The trees were dead, but the timber, except for its outside skin, could still be used. Throughout the 1940s a massive salvage operation was organised by the Forests Commission to harvest the dead trees before the timber rotted on the stump. It was urgent business. Foresters first thought that the fire-killed trees would deteriorate within two years.

The Black Friday fire therefore paradoxically gave new, if brief, life to many of the established timber-producing districts. It was the final fling of the tramway era. The principal areas of salvage were the ash forests. Their regeneration was expected to require special rehabilitation work, and their timber was the most valuable. Salvage operations occurred principally in the East and West Tanjil valleys, in the Thomson Valley, on the Yarra–Thomson divide, and around the townships of Rubicon, Matlock and Toorongo. Sawmill owners quickly established new mills in the burnt-out mountain forests. By the end of 1939, 48 new sawmills were salvaging fire-killed timber, and at the end of 1941 there were 68. By 1944 this had increased to over 120 and accounted for almost half of state forest sawlog output. Because of the emphasis on salvage work the largest concentration of Victorian sawmills remained in the ash forests, and few new areas were opened up. Petrol rationing and a lack of wartime labour also inhibited the relocation of sawmills. Forests that had already been logged were exploited beyond their limits.[43]

A dump of salvage logs in the early 1940s. Water sprays helped preserve the fire-killed timber.
(DNRE Library)

The salvage of fire-killed trees began to diminish by the end of the war, although it continued in many districts until the early 1950s. The timber had deteriorated less rapidly than expected. More than 1400 million super feet (3.3 million cubic metres) of fire-killed timber were salvaged in 15 years, half as much again as had been estimated by the Forests Commission in 1939.

One legacy of Black Friday was an increase in the number of large, fire-damaged trees which could provide nest-hollows for the rare Leadbeater's Possum. Leadbeater's Possum was thought to be extinct until rediscovered in regrowth ash stands near Marysville in April 1961. It had not been seen for 52 years. About an hour after dark in the Cumberland Valley on 3 April, members of a wildlife survey party noticed a small, grey possum low down on the trunk of a blackwood, its distinctive long, thin tail bushing out towards the tip. They took the first photographs of it alive. Leadbeater's Possum had been named prophetically after a taxidermist at Victoria's National Museum, for it was known for half a century only from a few preserved skins.[44] The possum is restricted to a forest of a particular age and fire history.

Such a precarious ecological niche illustrates the challenge of managing the ash forests for 'multiple use'. The possum depends on the existence of large, mature or dead trees. It nests in tree hollows, but the mountain ash does not begin to hollow until it is over 120 years old, way past the time when sawmillers consider it mature. Progressive and total clearfelling would exterminate the possum, as could another major fire in the next 100 years, but total protection of the forest from processes of renewal would not suit it either. The possum also feeds off young wattles. It was rediscovered in an area where it had access to nesting hollows in stags left by the 1939 fire, as well as good young feed from the regrowth. So Black Friday renewed the possum's habitat, but the very scale of it also created a monoculture of regrowth that will not provide nesting sites sufficiently quickly to replace the collapsing stags. Ecologist David Lindenmayer estimates that the loss of nest-sites will lead to a decline of at least 90 per cent of Leadbeater's possums over the next 15 to 50 years, and he notes that the population crash has begun.[45]

The conversion of tracts of old-growth mountain ash forest to younger stands of less than 80 years in age will also reduce populations of the Sooty Owl (*Tyto tenebricosa*) and Greater Glider (*Petauroides volans*) and possibly cause the loss of the Yellow-bellied Glider (*Petaurus australis*), all of which are dependent upon hollow-bearing trees. The presence of dead and burnt stag trees is also important for

Spencer's Skink (*Pseudemoia spenceri*, named after Sir Baldwin) which seeks out the sun and basks on tall, dead trunks sometimes up to 75 metres above the ground. The densest populations of the skink are therefore found in burnt or young regenerating forests, and even as the new forest soars skywards, the skink's arboreal skills on old stags enable it literally to rise above the dense canopy. Lyrebirds are found mostly in the younger stands. They cannot forage so successfully in the ferndominated older ash forests. Many birds and animals are adapted to, and attracted to, a forest in transition.[46]

Fire constituted one massive disturbance, and salvage operations among sensitive ash regrowth created another. Feet, machinery and rolling logs destroyed young seedlings. The balance between salvage work for immediate gain and restraint for long-term benefit was considered in many districts. A Powelltown forester reported in 1940: 'It is still a moot point as to whether it is better to salvage certain areas and in so doing destroy hundreds of seedlings or leave the timber to slowly rot and let the seedlings grow to form a new forest under conditions such as nature intended'.[47] By the end of the 1940s timber salvage operations were being curtailed in many fire-killed areas due to the increasing damage caused to advanced regrowth from logging operations.

Another threat loomed over the salvage work. The dead trees were 'matches' in a literal sense. Generally the fire threat to ash forests came from outside, but now it came from within. The annual report of the Forests Commission Fire Protection Officer for 1939–40 identified 'an entirely new and potent hazard' due to the 1939 fires.[48] Far from exhausting the fire potential of the region, the holocaust had created a different danger due to the dead ash trees and the debris falling from them. 'In contrast to a fire in a green Ash forest,' the officer continued, 'there is little danger to life in these fires.' The threat was to the ash seedlings, the hope of regeneration. A second burn of the seedlings would result in 'the permanent deforestation of [a] portion of the best water catchments – and their reversion to scrub areas.' He cited the 'denuded, eroded fire swept Toorongo plateau' (north of Tanjil Bren), which was burnt in 1926 and again in 1939 before the ash were old enough to carry seed, as providing 'evidence of the next step in the process of devastation.'

Fire detection and control were limited prior to 1939, and the methods of fighting were primitive. The practice was to use shovels and beaters, cut small fire breaks, burn back in a limited way, and use knapsack sprayers where water was available.[49]

Access to the forests depended on the narrow tramways, and there was a lack of heavy equipment to construct firebreaks. The 1932 fire had changed the opinions of Gembrook sawmiller Maurice Dyer about firefighting. Until he had witnessed a crown fire of such speed and ferocity he had thought that firebreaks and burning back were sufficient protective measures.[50] The 1939 fire gave that lesson to many more. It was a time for military metaphors. The Board of Works Engineer Alexander Kelso told the commission that to fight fire 'the type of organisation would be such as you would use if this country was invaded by an enemy – real centralised organisation.'[51]

The radical writer, H G Wells, visited Australia in January 1939 and was in Canberra at the time of the fires, attending a meeting of the Australian and New Zealand Association for the Advancement of Science (ANZAAS), an eminent gathering overwhelmed by smoke.[52] The heat was so sweltering that men were seen to remove their jackets and women their gloves, and 'the younger men of science went off firefighting'. Wells reported: 'The fires became at last such a preoccupation that there was nothing for it but to go up wind to the bases of that streaming smoke curtain and see the actual burning oneself.'[53] He found to his surprise that the fire front was unpredictable: 'A bush fire is not an orderly invader, but a guerrilla.'

He continually drew lessons from the fires for 'the coming war' and his essay about them developed into a polemic about how Australia had to defend itself from military attack in the same way it had to defend itself from fire: through aggressive prevention. 'Like these scorched homesteads we visited, [Australia's] real and effective protection lies in going to the source of the evil and beating it out there in time.' And he observed the psychology of fire, and of war: 'The thing to note, in a war-threatened world, is that for Australians, as for people at home, this sort of thing exhilarates. Everybody we met was dirty, hungry, thirsty, fraternal and quickened.'

The *Forests Act 1939* gave the Forests Commission responsibility for forest fire protection in all unoccupied Crown lands, and control of the use of fire on a strip of land one mile (1.6 kilometres) beyond state forests and national parks with the exception of the Mallee. This increased the commission's territorial fire responsibility threefold. The commission was also empowered to order the construction of dugouts. A delayed result of the 1939 experience, reinforced by fires in 1944, was the formation of the Country Fire Authority in 1945.

The Second World War brought advantages and disadvantages for the Victorian timber industry. On the one hand, there was increased demand for timber due to defence needs and the restriction of competition from imports. This led to a lowering of acceptable standards for products such as case timber, coming at just the right time for an industry geared up to rapidly salvage fire-killed trees. Young, dead ash, which might otherwise have gone to waste, was utilised down to a 7-inch (178 millimetre) diameter, about half the previously accepted size.[54] On the other hand, labour was scarce due to the war, and other forest activities – especially stand improvement – were curtailed. As a forest officer in Erica bluntly put it in 1941, 'Men are hard to get and are usually feeble and inefficient when obtained'.[55]

These three factors – salvage, fire and war – meant that in the decade after Black Friday the forests of ash were exploited intensively and urgently. Such high levels of output could not be sustained indefinitely. The increased production was achieved by intensifying sawmilling activity in existing locations, and parts of the ash forests were logged well past normal limits. A greater proportion of young trees, diseased and damaged trees, and 'rubbish species' were milled. Resources were taken away from silvicultural (tree cultivation) work. The area of Victorian state forest silviculturally treated fell dramatically in this period.[56]

In some senses the 1939 fires made forest managers urgently confront issues they would soon have had to consider anyway. Nearly every annual report from the Upper Yarra during the 1930s warned that the district was nearly cut out. A 1940 Forests Commission report from the Broadford district commented that 'There remain very few areas of virgin timber in the Broadford District'.[57] There was a need, urged the local forester, to develop uses for inferior timber while the first-class sawmilling forests advanced to maturity again.

Foresters encouraged the installation of 'case benches' in sawmills (for the production of timber for packing cases) as a way of making use of lower grade timbers. Increasingly poorer types of logs were taken from previously cut-over areas 'to maintain production at any price'.[58] Charcoal production, which was in demand as an alternative fuel while petrol was rationed, was urged as the best way to use such timber as well as dead and ringbarked trees.

In 1941 a Kurth kiln was constructed at Gembrook – the only charcoal kiln built in Victoria that could operate continuously. Emergency supplies of firewood were also required from the forests for military and civilian heating and cooking and as a substitute for coal for locomotives. In the wartime absence of labour, internee and prisoner-of-war camps were established in the forests to cut firewood.

From the early years of the Forests Commission, encouragement was given to the production of paper pulp from mountain ash. Ash species, with their short fibres and light colouring, were well-suited to the production of wood pulp, and the large-scale power logging of the mountain forests left considerable quantities of lower quality timber that was useless for sawing. In 1936 the government, the commission and Australian Paper Manufacturers Ltd reached an agreement that led to the establishment of a pulp plant at Maryvale in Gippsland. It commenced production in October 1939 and for several years drew its wood supply from the fire-killed ash forests.[59]

In the Erica district a marked shift from mountain ash to mixed species sawmilling took place in the decade after Black Friday. The slopes of Mount Baw Baw were being cut out rapidly before 1939, and this exhaustion was accelerated by the salvage work. By 1950 there was only one sawmiller in the district (C H Tutton Pty Ltd) operating solely on mountain ash. 'During the past ten years', wrote the district forester in 1950, 'another class of produce has gradually assumed prominence, namely pulpwood and this is now our second largest source of royalty'.[60]

The need to log new forests became urgent as the salvage work diminished. From the 1950s sawmillers began to re-establish themselves in East Gippsland and northeastern Victoria, close to Victoria's remaining virgin forests. Through the building of more roads into forests and the introduction of a Royalty Equation System in 1950, the Forests Commission eliminated some of the penalties of remote location and assisted decentralisation of the sawmilling industry. The Ezard family of sawmillers illustrated this staggered eastward movement in search of new, unexploited stands of timber. They began in the Warburton–Powelltown area in the 1920s, moved east to the Erica district in the 1930s, and further east to Swifts Creek in the 1950s. As long as virgin native forests were available for utilisation elsewhere, hardwood sawmillers were under little pressure to manage and work regrowth or plantation forests. It was easier to move to new country than to wait 80 or 100 years for renewal. In 1953 log allocations in the heart of the ash forests were cut drastically, and the focus of Victoria's sawmilling industry moved eastward.[61]

After the flurry of salvage work, people withdrew from the bush. Black Friday depopulated the forests of ash. Such was the danger of isolated bush life that sawmills were henceforth more safely established in towns. Roads took over from tramlines, and crawler tractors from winches; the chainsaw replaced the axe and crosscut saw;

diesel and electricity replaced steam power; sawmilling equipment became more sophisticated. The timber tramway era ended. Old forest workers who later visited the bush were, like the Walhalla woodcutter Lou de Prada, appalled by the silence. They missed the hoot of the winch whistle and found today's forests 'deserted and forlorn'.[62]

The effects of the 1939 fire remain. Gaunt, dead trees from the fire still rear above the crown of the regenerating forest. It was not just the fire but also its aftermath of intensive exploitation that shadowed a generation. The Forests Commission foresaw the problem in 1941. 'Current concentrated exploitation', warned the annual report, 'will call for a specially sustained post-war effort to restore forest equilibrium.'[63] Fire creates as well as destroys, and the post-war preoccupation was with the new forests to which Black Friday gave birth.

LEADBEATER'S POSSUM

JOAN M DIXON

Leadbeater's Possum is the only endemic possum known from Victoria and it is one of the state's faunal emblems. In 1867, when Victoria's forested areas were much more extensive than they are now, the first specimens of the possum, a male and female, came in from the heavily wooded Bass River Valley of South Gippsland. The Director of the National Museum of Victoria, Professor Frederick McCoy, described *Gymnobelideus leadbeateri* in that year, and wrote of the species: 'I name it after the skilful taxidermist [John Leadbeater] of our Public Museum, in which specimens of both sexes are preserved.' McCoy's description was accompanied by black and white illustrations of the animal, its jaws and teeth, by lithographer J Bazire.

In the *Prodromus of the Zoology of Victoria* (1885), McCoy repeated much of his description and the male was depicted in colour, accompanied by line drawings of the skull by Dr Wild, artist and lithographer.

Although the species was not considered rare in McCoy's time, numbers soon dwindled and few museum specimens were acquired in the next 50 years. Mr A Coles obtained a specimen from the Bass River District in 1899, and in 1915 a specimen was donated to the museum by Mr F V Mason. He noted that it had been taken many years ago 'from the edge of the Koo-Wee-Rup Swamp about three miles south of Tynong Railway Station. As a tree was felled, the little animal emerged from a hollow branch.'

Much later, a previously overlooked specimen from Mt Wills in north-east Victoria was found in the museum with other possum species. Records indicate that it was shot on the roof of an abandoned gold mine at Sunnyside '4000 ft' above sea level around 1909. In 1960, another old specimen located at the Burke Museum, Beechworth, also in the north-east, was sent to Melbourne. Although it lacked a clear provenance, it nonetheless reinforced the likelihood of occurrence of the species in the mountains east and north of Melbourne.

Fruitless searching for live specimens continued until early in 1961, when Eric Wilkinson, a museum staff member undertaking a mammal survey near Marysville, about 80 kilometres north-east of Melbourne, saw two animals which seemed to be Leadbeater's possums. In April of that year a specimen was obtained for the museum collection. This colony was living in mountain ash habitat, with a scrubby acacia understorey.

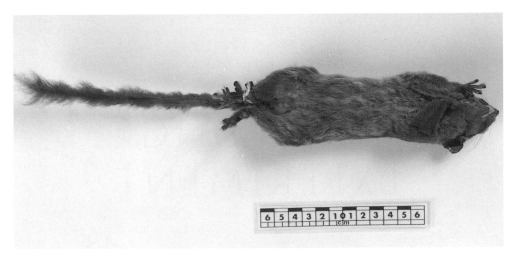

A specimen of Leadbeater's Possum from the Bass River area, Westernport 1867. (Museum Victoria)

Leadbeater's Possum is still regarded as endangered and is confined mainly to the western limit of Victoria's Central Highlands. A small isolated population is known from the Yellingbo area near Gembrook. About 70 per cent of the total population of this unique possum occurs in state forest, where timber harvesting is allowed; the remaining 30 per cent includes the Yarra Ranges National Park. Unevenly aged ash forests, where wattles and old hollow trees for nesting are abundant, are the optimum habitat. Threats to the species' survival include the gradual loss of these old trees by wildfire and timber harvesting.

11

THE LONG EXPERIMENT

Silviculture means 'cultivation of trees'. It is the science and economics of the establishment, management, composition and growth of trees. Silviculture is like farming; the forest is a crop to be harvested. Following Black Friday and its aftermath the managers of the ash forests tentatively commenced a long experiment, one that had been delayed by the pressures of utilisation but that now became urgent. How were they to foster regeneration of the mountain ash? How would they manage the young regrowth stands? This was the new silvicultural challenge.

The profession of forest management in Australia had been founded on a different silvicultural problem. Victoria's first Conservator of forests, George Perrin, continually emphasised that colonists needed all 'the old timber out'. He had 'no intention of keeping a single matured tree from the saw-miller.'[1] In 1918 the newly formed Forests Commission administrated what it called 'large areas of mature and over-mature forest which needed to be logged and replaced by young healthy stands'.[2] The forests they were to manage were still regarded as largely 'virgin'. In fact, they were considered 'over-ripe'. In 1937, not long before the Black Friday holocaust, the Chairman of the Forests Commission of Victoria reflected that 'at the present time all millable Ash stands are virgin forest – the oldest regrowth is not more than 25–30 years old … From both the silvicultural and economic points of view', he explained, 'these stands require to be exploited in as short a space of time as possible'.[3] They needed to be replaced by a more efficient forest, one of different-aged stands that could be managed and utilised continuously. This was the foresters' mission. There was urgency in their work even before the fires and salvage operations. The fires accelerated the urgency; the war excused it.

However, repeated large-scale fires also confronted foresters with a new problem. 'Over-ripe' and heavily logged forest had been replaced by hectares of even-aged seedlings and, in places, hillsides of bracken. In some areas the regrowth was promising; in others there was none. Further, it was not just frequent fire that spirited away an ash forest. The frequency of fires up to 1939 hindered until the early 1950s proper recognition of the general failure of *logged* ash stands to regenerate satisfactorily in the absence of fire. Foresters knew little about the mountain ash and the best conditions for its regeneration. They knew little about managing regrowth ash forests. There was, as forester A H Beetham put it, a 'long period of silvicultural uncertainty ahead'.[4] Victorians faced the immense anxiety of watching this economically critical forest renew itself, knowing that they would not themselves see the culmination. It was going to be a long experiment. They would not live long enough.

The uncertainty of mountain ash regeneration had been noted with concern by the earliest forest managers. In 1874 W E Ivey commented on nature's capricious cooperation. Although mountain ash was regenerating well after the 1851 fire (although the thick regrowth needed thinning, he observed), it was not renewing

Tubed stock of E. regnans *ready for planting out.* (DNRE Library)

itself after utilisation by splitters and sawmillers. 'Here seedlings do not, as a rule, take the place of the felled trees; but the forest in these places is left open, with an appearance of bareness.' It was in these places, he concluded, that nature needed help.[5]

Systematic assessment of the ash forests started about 1928 and continued throughout the 1930s. Mapping the complex mosaic created by sawmilling and bushfire in wet, rugged country was demanding work and confronted foresters with 'an acute appreciation of the lack of definite knowledge ... [about] the silvics of our indigenous species'. They surveyed parallel lines through the forests at 10-chain (200-metre) intervals and recorded types of forest, density of scrub, rock formation, occurrence of regrowth and diameters of the merchantable timber trees. In the Rubicon Forest Working Plan, for example, stands of trees were allotted to the categories of 'overmatured', 'matured', 'saplings' or 'seedlings'. Along the assessment lines were occasional sample plots where the diameter of every tree was measured, heights recorded, and the number of log lengths and their volume estimated.[6]

Following the fires of 1926, as with 1851, a dense regeneration appeared in many parts of the ash forests. To assist nature, foresters commenced ringbarking the remnant over-mature trees to ensure that their new crowns did not 'suppress' the new growth and to induce a heavy seed throw similar to that produced by fire. In the Upper Yarra district, 800 to 1200 hectares of forest were treated in this way each year in the late 1920s and 1930s. It was mainly mixed species forest but included some areas of mountain ash. The practice was not very successful and, they lamented, was wasteful of good timber from the mature trees.[7]

Much of this work was done by depression relief-workers and in some cases their lack of enthusiasm for the job may have been of long-term benefit. In the Mumbulla State Forest near Bega in New South Wales, unenthusiastic relief workers ringbarked only the sides of trees visible to their supervisors. Many of these trees now survive and provide valuable animal habitats.[8] It is possible that there are also some 'relieved' possums in the ash forests.

In 1932 another fire swept over new areas and some that had been burnt in 1926. Foresters noticed that the regeneration of mountain ash was patchy and asked the commission to undertake more studies of its flowering and seeding habits. In Britannia Creek Valley between Powelltown and Warburton, officers reported that 'regeneration has only occurred in isolated patches, the remainder being bracken waste'. In 1936 R Powles began experimental work on the artificial seeding of mountain ash in these bracken areas and found that no seedlings grew in areas sown without prior burning. Experimental thinnings of young stands of ash were also carried out in the Little Yarra Valley, and in the Ada River forest a study was begun to determine the best method of obtaining regeneration of mountain ash forests after logging operations.[9]

The 1939 fires, which occurred so shortly after the 1926 and 1932 fires and burnt many of the same areas, made it even more urgent to understand the circumstances

Dense scrub replaced ash forest after successive fires on this slope of Mount Observation in the Armstrong Creek watershed. Skeletons of the former forest rear above the scrub and occasional silver wattle. This photograph was taken by Jack Gillespie about 1960.

under which mountain ash successfully regenerated, but it also delayed that essential silvicultural work. More areas of promising young regrowth were devastated. In March 1939 a departmental survey of forests from Toolangi to Erica ascertained that 73 000 acres (30 000 hectares) of burned ash areas required artificial reforestation. District foresters continued to ask Head Office for more silvicultural advice. Why was regrowth so patchy and sometimes completely absent? How long before they would know whether it was established? How should they manage the 1939 regrowth? How many years would pass before it was ready to mill, and how would they maintain production in the meantime?

It was not until the late 1940s that foresters began to pick up the threads of silvicultural work again. A H Beetham, who had endured Black Friday in the Mansfield district, studied areas of regeneration in the upper Latrobe Valley. He urged the collection of information about mountain forests and felt that foresters had a fundamental role to play in their improvement. Untreated areas of forest, he concluded, 'continued to develop as wild unkempt and fire menaced tracts of bush', while those artificially improved were 'clean, accessible and protected areas of woodland'.[10]

In 1962 Jack Gillespie completed a Diploma of Forestry thesis on the denuded forest lands of the Upper Yarra Forest District where he was working as a forester, based at Powelltown. He noted that the virgin forests of large trees had generally been replaced by stands of regrowth. However, recurring fires had left denuded areas (one-tenth of the district's forest land) that were totally unproductive, where bracken, wattle and scrub species occupied the site. The majority of the 7000 hectares of denuded land in or very close to the Upper Yarra district studied by Gillespie had previously carried mountain ash. Some areas had been burnt three times in the 13 years between 1926 and 1939, and further fires in 1948, 1951, 1955, 1957 and 1958 had burnt areas already partially denuded.[11]

It was not just a question of regeneration after fire or logging. What of the remnant stands of old growth forest? Would they renew themselves? Could the ash forests regenerate without fire?

It was this question that in 1949 launched Melbourne botanist, David Ashton, on *his* long experiment.[12] Like many of Victoria's great naturalists, Ashton grew up in Melbourne's eastern suburbs and found his early interest in plants stimulated by the easy proximity of suburban bush paddocks. His botanical curiosity came alive in the most accessible part of the ash forests, the Dandenongs. He knew them well as a child, but when he visited them in 1946 on a Melbourne University Botany excursion, 'all of a sudden everything [he] saw was for a reason; they weren't just trees, they were all different trees, [and] certain trees were here and not there'. It was ecology he wanted to study, a science that would enable him to puzzle over whole living systems and their attachment to locale, why 'certain trees were here and not there'.

The tree he went on to study was *Eucalyptus regnans*, the mountain ash. The head of the Melbourne University Botany School, Professor John Turner, set Ashton on his task. Turner was an Englishman who had arrived in Victoria to take up his academic post in August 1938, just months before the Black Friday holocaust. His eagerness for field work and his desire to discover his new country ensured that Turner saw the mountain ash forests before they went up in smoke.[13] This brief reconnaissance allowed him to share in Victorians' sense of loss and sharpened his awareness of the scientific and economic challenge of ash regeneration. By the late 1940s he and the

Professor of Agriculture, Sam Wadham, together with the Chairman of the Board of Works, John Jessop, knew of a forest 'ripe' for study. Ripe, but not regenerating – why?

On the furthest western margin of the forests of ash, less than 50 kilometres north of Melbourne in the Wallaby Creek water catchment area, stands an ancient forest of mountain ash. Most of the Big Ash forest probably arose following a severe fire about 1730, and some from the Black Thursday fire of 1851. Hume and Hovell passed through it in 1824, paling splitters raided it in mid-century, water resource managers protected it in 1872, bullockies burnt its understorey repeatedly in the early twentieth century,

Profile diagram drawn by botanist David Ashton showing a mountain ash stand on the fringes of the Wallaby Creek Big Ash forest. It takes 2 or 3 field days to prepare accurately such a profile. (Courtesy of David Ashton)

but it was 1949 before anyone really looked at it. David Ashton wished then that those earlier observers had recorded more of what they saw, and he yearned to return to 1850 with a botanist's eye. Like every good ecologist, Ashton is a historian too.

He set about creating regeneration plots in the midst of the mature forest. He rode along the bracken firebreaks on his push-bike, found a natural gap in the mature forest canopy where trees had fallen, cleared the scrub, sowed mountain ash seed – some in the gap, some under the canopy of the mature forest, and some under bracken – and fenced in his 15 plots with wire taken from old pine plantations. In spite of the bloodsucking leeches and hard, lonely work, Ashton loved the big forest; it was 'an awesome but beautiful place' towards which he came to feel 'an animal territoriality' and 'a feeling of belonging'.

It was to be a three- or four-year project, he thought, when he sowed the first seed in late 1949. After five or six years, the seedlings in the plots under the forest canopy and those under the bracken had well and truly died while the plants sown in the gap were three metres high and growing vigorously. Ashton felt that he had the answers. Open a gap in the forest canopy of 1300–1400 square metres, clear the understorey, protect the seedlings from wallabies and regeneration was possible. 'But', he recalled, 'it was the next five or six, the next ten, the next fifteen years which just turned the whole thing inside out ... If I'd stopped then I would have been absolutely wrong.'

The vigorous young trees in the gap stopped growing, and the understorey species took over. Why? To answer that question, Ashton had to research the whole biology of the plant, its 'autecology', every single thing about the organism and its relations with its environment that he could discover. He 'began to pry into the personal life of *E. regnans.*' How fast did it grow? How deep are its roots at different stages? How often does it flower? How viable are the seeds? What happens to them? The secrets of the forest – the key to its future – were to be unearthed by this myriad of questions about the life of one organism.

Ashton pieced together the story with patience and care. First there is the 'miracle of timing', by which seed is released but not destroyed in a crown fire. The fire opens up the canopy to the sun. It also prepares and dries the soil, making it especially fertile so that the seeds can flourish. Ants, however, scurry off with many of them. When seedlings raise their heads, wallabies and wombats munch them off. More are lost to fungal attacks. Eucalyptus seedlings carry a heavy parasitic burden, particularly in cool, wet forests, and overcome it only by growing fast – in the case of the mountain ash, very fast. It can reach half its final height in 25 or 30 years, but to do that it needs everything going for it: plenty of water, nutrients and light. The light is crucial. As the seedlings grow, their horizontal leaves turn vertical. How does the sun reach them if they are shadowed on the sides, if the gap in the forest is too small? Trees 60 to 90 metres high cast long shadows, and vertical leaves intercept as well as crave light from the side.[14]

Ashton's work defined the main requirements for successful seedling establishment as the presence of an aptly named 'ash-bed' that is made receptive to seeds by fire, an adequate seed supply with discouragement of insects, and plenty of light. Mountain ash forests perversely need a catastrophe to survive; they have made a Faustian bargain with fire. They need Black Fridays. Clearfelling, the favoured silvicultural system for ash, tries to mimic the favourable natural events. A gap is created, light comes in, the soil is fired and enriched, the seed is 'pelleted' (coated with insecticide) and sowed. It is a man-made disturbance of massive proportions, and it seems that it needs to be so.

What will become of the Big Ash forest at Wallaby Creek? 'Although it has escaped fire for nearly a century', reflects Ashton, 'this seems to have been pure chance.' It is unlikely therefore that the skulking rainforest will replace it. The Hume Range is adjacent to drier foothills and plains to the north, west and south, places where fires have originated in the past. In 1982 a severe fire roared to within one kilometre of the Big Ash forest before it was suddenly deflected by a gale-force wind change. If the change had arrived just 30 minutes later, Ashton's experiment would have included the earliest stages of forest recovery.[15]

David Ashton and 'Mr Jessop'. Ashton named this tree – one of the tallest living trees in Victoria – after the Chairman of the Board of Works (1940–55) for his role in encouraging old growth forest research.
(Photo: Tom Griffiths)

In the post-war years, interventionist forest management received increasing scientific and practical support. Stretton's 1939 report had sanctioned controlled burning as a protective measure, and Ashton's research revealed the fundamental role of fire, light and disturbance in ash regeneration. In the same period, the ecological analysis and empathy that informed work such as Ashton's also underpinned a broader change in popular attitudes to nature conservation that would conflict with disturbance management. Forests were increasingly appreciated not just for their obvious beauty or grandeur, not just for their conspicuous tall trees, but also for the myriad of less visible life forms they embowered. They were valued for their secrets. Forests were so dense and rich and mysterious that they alone perhaps could harbour a non-human world. They were primeval; they were wilderness. Such values would be destroyed by interventionist forest management. What would controlled burning do to biodiversity? How could clearfelling – the brutal, revealing 'hand of man' – leave any secrets?

Foresters were peeved by the modern environmental movement and shocked to find themselves the subject of so much of its criticism. After all, they were (they kept saying) the first conservationists.[16] They had been one of the few restraining influences on the use of the forests. They had fought for the preservation and scientific management of timber resources against the political priority of clearing and settling. Applied scientists, such as foresters, hydrologists and soil scientists, were the earliest advocates of 'conservation', by which they meant the 'wise use' of natural resources.[17]

Mountain ash sawmiller Hec Ingram spoke to foresters in 1979 of his fears for their joint business. He passionately urged foresters to speak out, to defend themselves and their terrain. '[Y]ou are one of the few professions, and could be the only profession, that has the full confidence of the public', he told his audience. 'But', he warned, 'there are groups of people, misguided I think, who would enter your domain.' It was indeed partly a battle over domain, over bureaucratic power. Just as the Forests Commission had been suspected of criticising cattlemen in order to wrest control over Crown Lands from the Lands Department, so now did the Forests Commission find itself losing forest land to 'even better conservationists', the National Parks Authority (later Service), formed in Victoria in 1956.

John Youl, a forester involved in the post-war salvage and firewood operations, spoke at the same function as Ingram and listed the notable changes in sawmilling that had taken place in his lifetime. 'Number 13', he concluded, 'probably aptly numbered, is the sudden emergence and build up in the 1970s of the influence of the conservation and environmental lobby.'[18] Many foresters pictured it as an event as sudden and catastrophic as Black Friday. They were taken unawares, and felt betrayed and misrepresented.

From its beginnings, the Forests Commission had a role in public education, recreation and conservation, going beyond a purely economic valuation of the forest resource. In the three years to 1922, the commission issued 83 520 trees to state schools (free) and 148 115 to the general public, in an effort to encourage tree planting.[19] Building a public 'forest conscience' had long been one of the aims of the commission, and this goal was given further vigour by the Black Friday fires and the formation of a 'Save the Forests Campaign' in 1944.[20]

The commission also fostered forest recreation, although this role was not formally recognised until the *Forest Act 1958*. In 1950 the entire Monbulk State

Forest and part of the Sherbrooke Forest were combined to form the Sherbrooke Forest Park, and timber production ceased. At the same time, following representations from groups such as the Field Naturalists' Club of Victoria, the Commission recognised the need to establish wildflower sanctuaries within some of its forests. 'Multiple use' and 'sustained yield' of the forests were aims frequently stated, if rarely pursued in practice.

Utilitarian and romantic views of the forests went happily together. The Australian Forest League, which began in Victoria in 1912, championed the professional management and use of forests, and its journal, *The Gum Tree*, became the official organ of the Federation of Victorian Walking Clubs and the League of Treelovers. It was subtitled 'A paper for all Bush Lovers'. The 'Save the Forests Campaign' counted senior forestry professionals and industrialists among its leaders; people such as Charles Ewart of the Forests Commission and Sir Herbert Gepp of Australian Paper Manufacturers.[21] 'Wander with us among these majestic beauties', urged a Forests Commission film of the 1930s. 'The sap that runs through our trees may be termed the lifeblood of the country.' Against images of toppling giants, the narrator enjoined us to take care of our forests – 'a great heritage' – which, it was explained, was now safely in the hands of trained scientists, the foresters. These new managers boasted of 'an organised destruction'.[22]

That phrase summed up the compromise that was at the heart of the foresters' business: they were both the facilitators and regulators of private interest. Their work aimed for maximum timber production from the resource in their charge. Forestry stood for a different sort of 'improving'. It was another type of farming, and silviculture another way of making 'waste' land productive. Wombats and wallabies, just like rabbits, were considered 'vermin', and foresters' district annual reports detailed eradication work under this heading. In this sense, forestry was not in tension with farming, mining and settlement; it could be seen as just another form of civilising, another imported tradition contemptuous of indigenous natural values.

Stephen J. Pyne has described foresters as 'among those transnational Europeans of the nineteenth century' who carried their civilising expertise around the world.[23] Australian forestry was certainly shaped by the notion of 'empire forestry' with its visiting experts and regular imperial conferences.[24] In the forester's hands, Australia's forests were no longer wild, primitive obstacles to civilisation, but agents of colonisation. Foresters, then, became prime targets for an environmental movement that emerged in the late 1960s at a time of resurgent nationalism and in a period when that other bogey – clearing for agricultural settlement – was diminishing.

Forestry's priority for the wood production values of a forest became clearer in the post-war period as areas of virgin forest diminished. The high demand for timber led to continued neglect for silvicultural work. The sawmilling industry, once largely in the hands of local family firms, became centralised into larger corporations, and bulldozers, chainsaws and road trucks revolutionised its destructive capacity. In the 1960s the clearfelling of Australian forests for the Japanese woodchip industry (particularly on the southern coast of New South Wales, in the south-west of Western Australia and in Tasmania) offered foresters the opportunity for total wood utilisation. It was a sensible aim for the timber industry, and foresters had long encouraged the use of all parts of a felled tree, but now the aim went mad as thousands of hectares of eucalypts were razed and sold cheaply overseas. To cater for a demand for fast-growing softwood timbers, other areas of native forest were removed and replaced by plantings of exotic pines. This was imperial forestry indeed; local interests, indigenous timbers and natural flora and fauna were being sacrificed to overseas markets and exotic plantations.[25]

In 1996 the Central Highlands which embrace these forests of ash were assessed as one of the Regional Forest Agreement areas identified by state and Commonwealth governments. Of the 908 000 cubic metres of native forest wood logged per annum in the Central Highlands, 72 per cent was chipped for paper production either in Australia or overseas. Sawn timber therefore accounts for only a third of the resource. History – with its eye for locality, community and continuity – can easily overlook this startling reality.[26]

But an increasing regard for natural forest values has also continued to revolutionise policy, as is apparent in the evolution of the concept of 'control burning'. Stretton's report of 1939 had recognised the need for more protective burning of the forests, by which he meant strip burning of firebreaks and patch burning of small areas for fuel reduction. The protection of people, property and timber resources was paramount. In the 1960s the concept of loss due to fire broadened to include not only human life, buildings, fences, stock, pastures, crops, timber and water yield but also flora and fauna. The effect of control burning (which had come to mean low intensity fires over larger areas) on the non-wood values of forests began to be investigated.

The Forests Commission of Victoria noted in its 1964–65 annual report that studies of the effects of fire on timber quality had been broadened to include the effect of prescribed burning on water catchment values and bird populations, and later the commission sponsored research on the effect of fire on certain animal species and wildflowers. There were, of course, different types of fire, and foresters began to accept that the fire regime best suited to the protection of timber and humans might not also be the best for plants and animals. Also, fire regimes would be varied and local. In the words of Professor John Turner, a 'new field of applied ecology' had emerged.[27] Should control burning be used in national parks? Should it be used, or allowed to happen naturally, in designated 'wilderness' areas? 'Applied ecology' swiftly became 'applied history', for knowledge of the past use of forests and fire became essential to determining whether modern control burning was recovering a historic landscape or creating something entirely new.[28]

The practice of clearfelling also poses historical questions. Does it accurately mimic natural fire? Is clearfelling good history? Just as Black Friday stimulated post-fire ecological science, so did Ash Wednesday 1983 seed a further generation of research. On 16 February that year, ahead of an intensifying double cold front that hurled hot northerly winds from the inland before it, fire went on a 12-hour rampage across South Australia and Victoria. Tornadoes of flame exploded 375 metres into the sky and the hot breath of disaster shrivelling the mainland could be smelt on the south-west coast of Tasmania. Whereas Black Friday found its origins in rural burning, Ash Wednesday was entwined with the fate of the urban bush. It seemed to melt the distinctions between city and country and claimed lives on the suburban fringe: some of the most devastated areas were in the foothills of the Dandenongs. With a shocking symmetry, 71 lives were lost in this great fire, too.[29]

Peter Attiwill was one of the scientists who entered the forests while they were still smoking and closely monitored regrowth after the fires. He has used the forests of ash to explore ecological paradigms as well as management strategies for Australia's eucalypt forests. Like many late-twentieth-century scientists, Attiwill rejected Frederic Clements' older ecological idea of a climax community. He saw the mountain ash forests – a community that depends on catastrophic disturbance for its perpetuation – as the ideal place to elucidate an alternative ecology of disturbance.[30] Clements' model was long ago discredited among ecologists, but Attiwill felt that its hold on the public mind remained too strong. 'Disturbance', he argued, 'is a natural and prevalent feature of ecosystems.' In the forests of ash, 'there is no stable and self-perpetuating end-point' and 'maximum diversity and productivity is maintained by random periodic disturbance.'

Attiwill grew up listening to Crosbie Morrison's nature talks on the radio, and also went on many bushwalks with him, because Morrison was a friend of his father's. Trained initially as a forester, Attiwill completed a doctorate in soil fertility and plant nutrition, and joined the School of Botany at Melbourne University in 1966.[31] His biochemical focus made him interested in the way an ash forest recycles soil nutrients. Would a massive disturbance such as Ash Wednesday cause a loss in soil nutrients, either through leaching or by being turned to gas in the intense heat? The 'astoundingly rapid' regeneration rate of the trees he studied at Britannia Creek – which he found to be 'the world's most productive ecosystem' – convinced him that this did not happen, and that stand-replacing fires create the right conditions for the conservation of nutrients and their uptake by regenerating seedlings. On the basis of his monitoring of levels of soil nutrients and carbon storage, as well as plant diversity and productivity after intense fire, Attiwill has argued the ecological equivalence of clearfelling and natural large-scale disturbance. Logged forest and burnt forest were comparable. He concluded that clearfelling ash forests is an ecologically sustainable form of management.[32]

Attiwill's career illustrates the uneasy relationship that has developed between ecology and politics, especially since the 1960s. In 1969 he gave evidence to a Select Committee of the Victorian Parliament to inquire into the proposal to develop the Little Desert in the dry north-west of the state. He spoke strongly for the preservation of all remaining unalienated areas in the Little Desert on the basis of his scientific assessment of the region's 'biological diversity', then a new term in public debate. He also spoke of 'a new morality', which he described as the need to maximise the quality of life in a finite and overpopulated world. It was an unambiguously political statement which he added to his scientific testimony. Yet over 20 years later, and in the midst of 1990s forest conservation politics, he denied the political dimensions of his Little Desert intervention, saying: 'I think my role in the Little Desert dispute was as a scientist, not as a political lobbyist.'[33]

Attiwill felt that the conservation movement 'became political' in the early 1970s. In the 1990s, as a controversial scientific advocate of clearfelling, he continued to distinguish between 'pure science' and 'advocacy', and to lament the 'weakness' and 'fuzziness' of ecology which meant that his science was denied 'the opportunity to be seen as an impartial arbitrator in public debate on environmental issues'.[34] 'It's okay for the Wilderness Society to argue on spiritual or philosophical grounds that you shouldn't log the forest', explained Attiwill. 'But the science says you can.'[35] Some conservationists have attacked Attiwill's scientific integrity and depicted him as a lackey of the forestry industry.[36]

Attiwill, the son of a former chief of staff at the Melbourne *Argus*, and Graeme O'Neill, a science writer and journalist, criticised the media's polarised coverage of environmental issues. In forest debates, they argued, one was expected either to be nature-centred and apocalyptic or an advocate for employment and the economy. Public debate allowed for no middle ground between nature and culture, ecology and economy. Attiwill and O'Neill believed that the green movement exacerbated this dichotomy by continuing to emphasise outmoded popular ecology that dictated 'on the one hand, undisturbed nature is stable, harmonious and unchanging, whereas, on the other, disturbed nature is fragile'. How, they wondered, could current ecological paradigms get into the political debate without 'the scientific messenger being shot'; how could one advocate a disturbance paradigm without being typecast as exploitative?

According to this analysis, if one argues for sustainable logging, as Attiwill does, then such a stance defies the preferred polarities of public debate and is therefore too subtle for fair reportage. And if one depicts an ecology of disturbance, then it offends the popular view of 'the balance of nature', and 'the messenger will be shot'. But Attiwill is no mere messenger for an uncontroversial science. His politics are that his ecology is not political. He discerns a political rejection of his 'pure science'. Yet there is a serious ecological critique of his work, and it takes disturbance very seriously and is more historical than biochemical.

David Lindenmayer is another notable ecologist who grew up in Melbourne's eastern suburbs. He was born in Balwyn, living there until he was 10, and he would have seen the clear blue outline of the Dandenongs and Mount Donna Buang from the end of his street. Lindenmayer's scientific introduction to the forests of ash was after Ash Wednesday 1983 'when Bob Hawke broke the drought'. He spent months working alone in heavy rain among the tall trees at Cambarville near Marysville, collecting data for Andrew Smith, a scientist researching the life of Leadbeater's Possum.

Lindenmayer has since spent the best part of two decades thinking about and working in the ash forests, and his interest in the fate of Leadbeater's Possum has propelled him into a passionate exploration of the changing structure and dynamics of the whole forest.[37] Whereas David Ashton's work began with a plant and focused on the long-term monitoring of selected sites of regeneration, Lindenmayer's research began with an animal, one that prefers a multi-aged forest, and he has studied hundreds of different sites in search of an understanding of the natural and historical variability of the habitat.

He believes that the integration of history and ecology has 'major potential implications for the way forests are managed and the conservation of the considerable biological diversity associated with such ecosystems'.[38] Is there really such a congruence between natural and human disturbance regimes? Lindenmayer argues that after clearfelling the forest recovers, but it is markedly different from what it replaces. Numbers of hollow trees, tree ferns and large decayed logs are severely depleted. 'Without changes to logging practices, many species in Mountain Ash forests have an uncertain future.'[39] His work profoundly challenges the ecological justification for clearfelling as it is currently practised.

In the immediate aftermath of Black Friday and salvage logging, the management priority was regeneration. How could these forests be renewed? Now, over 60 years after the defining fire, the urgent management question has become: how should we cut them down? 'How can we create young trees?' has become 'how can we keep old trees?' Scientific interest in the personal life of *Eucalyptus regnans* has led us to a scrutiny of the whole forest ecosystem, of all the animals and plants that inhabit the forests of ash, the canopy, understorey, leaf litter and great streamers of bark. The wide eyes of Leadbeater's Possum have become a symbol of that changing concern. Yet those eyes lead us back to the trees themselves, and to complex readings of forest structure. In search of the secret of regeneration and the life of the seedling, we needed to discover the awesome power of fire and light in this forest. But now, as we seek to manage a forest over six decades old, we are more ready to hear the evidence of the tree's mature durability.

Mountain ash trees are not always killed by wildfire and they do not always exist in even-aged stands. Intense stand-replacing conflagrations are not the only way that fire enters the forest. Low intensity ground-surface fires also occur. Some mountain ash trees do survive fires, even intense ones. Some trees bear the scars of at least seven different fires. A majority of trees survived in about 70 per cent of the area of mountain ash forest that was burnt in the 1983 fire. Furthermore, there is some historical evidence to suggest that multi-aged stands of ash (producing the mixture of young regrowth and older, living trees favoured by many animals) were more common in the past.

Although only about 10 per cent of mountain ash forest is multi-aged today, research by Lindenmayer and Mike McCarthy suggest that at least 20 to 30 per cent and up to 70 per cent of sites that are now even-aged, were multi-aged at the time they were burnt in the 1939 wildfires. And this is most likely a conservative estimate due to the fact that many old trees that survived fire did not survive the following two decades of salvage logging. Photographs and paintings from this region in the nineteenth and early twentieth centuries also show multi-aged forests and under-storeys that are younger than canopies.

David Ashton has aptly summarised the variation that existed in tall open forests: 'The big forests of both *E. diversicolor* and *E. regnans* possessed structurally and floristically variable understoreys at the time of discovery by Europeans. Persistent folk-lore relates both to the riding of a horse through the virgin forests and to the hacking of a path for progress … Such variation can be interpreted in terms of persistent, patchy, fire regimes'.

Black Friday has been a powerful template for clearfelling as a management strategy in these wet mountain ash forests. Clearfelling operations remove all merchantable timber and then burn the residue in a high intensity regeneration fire, thus creating an ash-bed for the growth of a new stand; this process is repeated on any given site every 50–120 years. However, in a wildfire, clearly some large mountain ash trees *do* survive and remain as living habitats, and a greater number of dead trees are left standing than is the case after clearfelling.

Lindenmayer's research has shown how the number of these habitats is sig-nificantly reduced in clearfelled stands in comparison to those that have been burnt. Furthermore, the sheer extent of Black Friday may have masked the previous prev-alence of multi-aged stands. The first half of the twentieth century produced a dramatic concertina effect of change in the forests of ash. The rapid succession of fires from the early 1900s to the 1930s meant that Black Friday burnt a human forest

INSECT OUTBREAKS

ALAN L YEN

The period of production forestry in the mountain ash forests of Victoria has seen some major insect outbreaks. There have been periodic major outbreaks recorded within the higher elevation eucalypt forests of south-eastern Australia since 1880, but the post-1939 outbreaks have been more carefully documented.

Didymuria violescens is one of the leaf-feeding stick insects (phasmatids) that chews the leaves of several different eucalypt species in south-eastern Australia with a preference for peppermints. The females drop their eggs to the ground, and the emergent nymphs have to find a food tree to climb. The eggs resemble plant seeds, and are harvested by ants and eaten by birds on the ground. The other natural enemies are egg parasites and birds that feed on larger nymphs and adults. In the mountain ash forests of Victoria, outbreaks of *D. violescens* were reported in the early 1960s to 1970s in even-aged forests that regenerated from the 1939 fires. Mountain ash was found to be more susceptible to the insect.

Why did these outbreaks occur? Research by F G Neumann and others suggested that the 23 years between the 1939 fires and the outbreaks of the early 1960s permitted the development of a deep litter layer which concealed phasmatid eggs from bird and ant predators. Further, the closed mountain ash canopy resulted in less favourable conditions for ants that collect seeds or predate upon young nymphs. Ants prefer a more open and lighter habitat. Another scientist, J L Readshaw, proposed a breakdown in control of phasmatid numbers by insectivorous birds, and also noted that outbreaks of *D. violescens* moved further south in each successive generation from northern NSW (in 1948) to Powelltown in Victoria (1960).

In 1984, a small sap-sucking species of insect was found attacking mountain ash regrowth of 1939. It was an insect new to science, a type of psyllid bug that was described in 1989 as *Cardiaspina bilobata*. Leaf and root pathogens are usually found associated with psyllid outbreaks, and it is possible that the psyllids may benefit from an increase in nitrogen levels in the mountain ash leaves as a result of the breakdown of plant proteins caused by fungal pathogens. The building of logging tracks and ground disturbance through forestry operations may have affected the hydrology of these forests, and could form a contributory factor in the build-up of fungal pathogens and psyllids. Museum Victoria undertook monitoring and life history studies of *C. bilobata* for the Department of Natural Resources and Environment

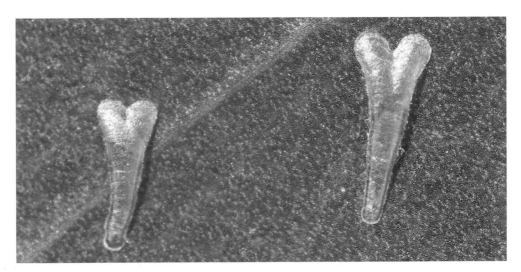

The lerps built by nymphs of the psyllid bug, Cardiaspina bilobata, *that sucks sap from the leaves of mountain ash.* (Museum Victoria)

from 1994 to 1998. Some mountain ash areas heavily infested in 1984 are now free of psyllids.

The phasmatid and psyllid outbreaks on mountain ash have occurred approximately every 20 years following the 1939 fires. In 1966 K G Campbell suggested that a single species eucalypt forest that is free of a catastrophe (e.g. fire) will have an outbreak of insects ordinarily in low numbers in a climax mixed-species forest every 20 years or so. How important is this in the life cycle of mountain ash – which can be 100 to 200 or more years? More interesting is the movement of *D. violescens* outbreaks south over time and the possibility that *C. bilobata* previously occurred further north. It is possible that both species were in lower numbers on other species of eucalypts in mixed species forests, and their dispersal to a single species forest enabled a release that led to outbreaks.

Mountain ash naturally occurs as a single species forest. It has survived at least two major insect outbreaks since 1939. What will happen over the next 20 years? Is it more vulnerable to outbreaks than eucalypt species that occur in mixed eucalypt forests and woodlands? As mountain ash trees are long-lived, how important is defoliation in their life cycle? Is a high level of leaf loss and reduced plant growth (whether natural or caused by human intervention) a reaction to particular environmental conditions? These are further dimensions of the long experiment.

legacy, and it was more ferocious as a consequence. Furthermore, the great fire's scale and timing produced unusually intense and sustained salvage logging which further simplified the forest structure.

Lindenmayer's conclusion, then, is that the structure of forest stands following fire is much more complex than many foresters have recognised traditionally or that current harvesting methods can accommodate.[40] Management of the ash requires a shift from 'the extensive use of clearfelling to the adoption of new silvicultural techniques that maintain more structurally complex multi-aged stands'.[41] There needs to be a finer scale of habitat conservation, resulting in the systematic retention of structural elements such as living trees, standing dead trees, logs and patches of understorey vegetation.

Lindenmayer would share Attiwill's frustration over the slippage between science and public discussion, and the simplifying habits of the media in its coverage of forest debates.[42] He is himself often identified primarily as a champion of Leadbeater's Possum, whereas his research is most significant for its landscape-scale ecology. However, Attiwill and Lindenmayer hold different views, not only about the practical management of ash, but also about science and certainty. With his wide-ranging data, statistics and simulation modelling, Lindenmayer exploits the 'fuzziness' of ecology and turns it into a science of probability, variation and historical contingency. History is not a template and ecology is not a blueprint.

The forests of ash again defy some of the general patterns of history and nature found in Australia's lower and drier eucalypt forests. Control burning hardly applies to forests of mountain ash, alpine ash and snowgum, and some modified version of clearfelling appears essential. The 1939 regrowth pushes skywards as foresters, sawmillers, ecologists and conservationists argue over definitions of 'maturity'. The long experiment remains unfinished.

12

HERITAGE

The forest grew back. It began to heal itself, enfolding the secrets of its recent history. Discovering the past has always been part of the excitement of living in or visiting the forest. South Gippsland selector, W H C Holmes, found that scrub-cutting was like a form of archaeology and that each blow of the axe penetrated an earlier layer of forest life. Settlers clearing their blocks found Aboriginal stone axes, blacksmiths' tools left by early surveyors, overgrown roads, forgotten mine shafts or simply the evidence of earlier axe-work. John Western discovered an old, rusty pick left by a prospector. Holmes himself remembers coming across the remains of yet another billiard table.[1]

When novelist Nettie Palmer wrote a book called *The Dandenongs* in 1953, she reflected on the special difficulties of 'breaking the hills', of 'penetrating the curtain that hides the past':

> There are few memorials there to call up the past, hardly a landmark to show
> the slow stages by which they have been won ... There are, it must be said,
> very few written records. It is only by burrowing into people's memories,
> exploring old sites, sifting various kinds of rubble, that some hint of the
> beginnings may be found.[2]

The entwining forest, the deep, rich leaf litter of the mountain ash and the wet soil all rapidly swallowed the traces of human occupation; fire regularly razed them, young dense regrowth screened them, blackberry thickets entangled them. In a land

of fire, custodians of culture came to rely heavily on memories, relics and an ability
to read the forest itself – the monumental biological record.

Historians of forest settlement echoed Nettie Palmer's sense of a precarious past.
For the first issue of the Mount Dandenong Historical Society Newsletter in 1974,
P J Hogan wrote:

> [H]ere in the Dandenongs emphasis has been on practical economy as settlers
> survived summer bushfire and the severity of rain forest winters, and we find
> we are inclined to have a diminished heritage, where the very fact of survival
> may be cause for veneration.[3]

Historic sites in the forest are rarely grand or beautiful. They may occasionally be
so, their very rarity cocooned by isolation, their delicacy a consequence of it. On the
old Jordan goldfields near Red Jacket, for instance, there survives an outstanding
example of quartz kilns, where rock was roasted before crushing by the stamp battery. The massive stonework erupts from the ferns like the ramparts of a castle. At
Noojee a trestle bridge spans a gorge, a graceful lacework of local materials. However,
forest relics are generally more casual and temporary – and if they were not actually
intended to be that way, the forest soon makes them so. Their aesthetic appeal comes
from the entanglement of culture and nature, the subsidence from one state into
another as the forest reclaims them. Their dignity is in their vernacular form, for
many relics were literally carved out of the trees. One of the attractions of forest
archaeology is discerning the 'hand of man' in the sculpture of nature.

Many historic sites are rubbish dumps; their 'treasures' were simply not worth
taking away. Mining and sawmilling were mobile industries; when a forest was cut
out or a reef exhausted, useful machinery was moved too. Some historic residue has
proved useful in unexpected ways. Sawdust, the waste product of sawmilling, which
remains in giant heaps at mill sites, is in demand by plant nurseries, and the more
historic the dust the better. It is also used in the making of lightweight bricks. Some
foresters with a sense of history have protected their district's sawdust heaps from
depredation. They regard them as signposts to the industrial past, and their size –
sometimes over 100 metres in diameter and 15 metres high – as a measure of milling
activity. Sawdust heaps can be seen on aerial photographs; they provide rare clearings
in the ash forests and resist revegetation.

Tree stumps are among the most evocative of historic monuments. Their meaning varies. On the settler's block they were statements of possession, even dwellings. In the forest, tall stumps could mean neglect. For a forester in 1887, 'tall stumps will long remain to remind us of the neglect which the forests have received'.[4] Their height represented waste timber and poor forest management. Near the rim of the stumps can still be seen the slots where the axemen stood on their boards to get above the broad buttresses. Or the stump may still be circled with cable, a clue to its role as an anchorage point in a winching operation. In some areas mountain ash stumps march at 10-metre intervals through stands of wattle, ghostly reminders of past giants and of a lost forest.

Some ancient trees remain, sometimes as isolated and celebrated specimens. In places the only big timber left in the forest is to be found in the remains of tramway bridges. The prostrate beams dwarf the living trees. Other clues to forest history abound, ready to be scuffed out of the leaf litter. "There must have been millions of six inch nails used in the timber industry', mused one Warburton resident,

Stumps of giant mountain ash recall the former forest around the Cambarville sawmill in the Cumberland Valley. Clearly visible are the slots where boards were secured for the axemen to climb above the buttresses. (Photo: Tom Griffiths)

'almost anything could be repaired or held together by them'.[5] These 'dogspikes', and the associated earthworks of tramway formations and sawmill pits, are among the hardiest of relics.

Bogie wheels also remain near the tramways along which they ran. Bridges, collapsed or leaning precariously, some held together only by the tramway rails, are among the most dramatic of forest structures. At sawmill and mining sites giant boilers loom out of the ferns. Stamp batteries, flywheels, waterwheel hubs, saw blades, cyclones (which separated sawdust from air), dugouts, cement kilns, fire-bricks and foundations all survive, a broken and incomplete jigsaw of clues. Collapsed planks, galvanised iron and brick and stone chimneys record the sites of huts. In some places, accommodation huts still stand. Their humble gables and remnant exotic gardens conjure a misleading idyll in the heart of the modern, uninhabited forest.

Some of the first raids on historic forest sites were made by bottle collectors. Societies of 'bottlectors' were formed in Victoria in the 1960s and 1970s, bottle

Often the only big timber left in the forest is found in historic structures. This huge cross-log supported a tramway to Paradise Plains near Marysville. (Photo: Peter Evans)

auctions were organised, competitions for the 'find of the month' were held, and meetings took place for members to show off their discoveries and arrange swaps. One member of the Melbourne Historical Bottle Society found a few Madeira bottles, 'some still with the original wine in them … In fact it was this that started me off on this hobby of bottle collecting', she confided. These groups had little regard for the provenance or context of their finds. It was the artefact they wanted, and they wanted it in *their* collection.[6]

Local history enthusiasts have grown in number in Australia since the 1960s, and they have congregated around collections of relics. Many of the new historical societies were formed to save buildings under threat or at least to safeguard more portable and collectable vestiges of the past. The communities of the forest have dispersed or retreated to the railheads, and it is around the railways – the declining symbols of their prosperity – that the historians of these communities have gathered.

The Upper Yarra Historical Society was formed in 1966, the year after the railway between Lilydale and Warburton closed, and it arranged to lease the old station building in Yarra Junction as a museum. The Alexandra Timber Tramway and Museum group occupies the Alexandra railway station and has a display that includes a large, circular, narrow gauge track with an original locomotive and tractor that worked the Rubicon forest. In the museum are working models of sawmills constructed by Ernie Le Brun, who worked in the Rubicon forest after 1927. Along the narrow gauge line between Belgrave and Emerald Lake in the Dandenongs runs Puffing Billy, a steam train that once worked the full line to Gembrook. In 1953 the line officially closed due to heavy financial losses and a landslide that covered the track, but Melburnians promptly formed the Puffing Billy Preservation Society to keep the train in operation. It survives as one of Victoria's greatest tourist attractions, carrying over 200 000 passengers annually. The society also runs a museum adjacent to the railway station at Menzies Creek and is currently restoring the railway to Gembrook.[7]

Railways lured another group of enthusiasts into the ash forests and made them, almost by accident, the most important researchers into the history of the forest. The Light Railway Research Society of Australia (LRRSA) was formed by five model-railway enthusiasts in Melbourne in 1961. For one of them, Frank Stamford, who had relations in Emerald, the closure of the Puffing Billy line when he was a child in 1953 was upsetting and formative. He later decided to find out everything he could about the Powelltown tramway, another railway in the forest. He and others in the society

were determined to bring a wider vision to their study of railways. They knew that
their technological interest could be narrowing: '[M]ost railway histories could just
as well be describing a railway on the moon', explained Stamford. However, it soon
seemed natural to broaden the story: 'When I was spending nearly every Monday
night for several years in the La Trobe Library going through old newspapers ...
I started to feel for the actual communities'.

So through the society's journal, *Light Railways*, members were encouraged to
research the social and economic context of the routes they studied and to provide
full references for their articles. They were urged to go into the field, to think of
themselves as 'industrial archaeologists', to interview sawmillers and foresters, and to
chase up old photographs, maps and government archives. They were following
Nettie Palmer's formula for 'breaking the hills': burrowing into people's memories,
exploring old sites, sifting various kinds of rubble'. As a result, the society rapidly
became the most prolific publisher of books on the Australian timber industry. It has
gathered a membership of over 500.[8]

The LRRSA focuses on research, not collection or preservation. So its celebration
of relics takes place in the field or via slide shows. After a trip to the Little Ada
tramway bridge in 1989, one member (Phil Rickard) wrote:

> Can anyone say that to stand in the chilly waters of the Little Ada and gaze
> up at this venerable bridge is any less an experience than to gaze up, at say,
> Flinders Street Station or St Pauls Cathedral? They are all monuments to
> their designers and the men who built them.[9]

This society of 'pious bush-bashers', as some called themselves, one day went in
search of 'the holy grail itself – the saddle tank from "Squirt" [a small locomotive]'.[10]
Members obviously enjoy poking fun at this semi-religious aspect of meetings and
excursions. At one meeting it was reported that:

> More than 40 Light Railway devotees congregated at the Ashburton Uniting
> Church Hall on April 14 [1988], for their bi-monthly fellowship pilgrimage
> ... After [the President] Bro. McCarthy had delivered his sermon, several of
> the congregation had their say – chapter and verse! ... [They then] took up
> the collection of offered slides and proceeded to project their images onto a
> white shroud.[11]

The research orientation of the society makes it an active and lively one; a meeting may be conducted on a sawdust heap, and a crowbar, axe and machete are useful on excursions.[12] Following the 1983 bushfire, members were involved in a frenzy of fieldwork as they hurried to map the newly revealed lines of old tramways. 'The 1983 bushfires were a godsend to us – it's terrible to say that … It was like a veil being removed'.[13] Heritage professionals have sometimes overlooked the careful survey work of the society, partly because they are regarded as 'amateurs' and partly, perhaps, because the society is ambivalent about preservation.[14]

The society has rarely involved itself directly in the conservation of forest relics, although some members have acted to protect and register sites. Perhaps such conservation is unrealistic, even inappropriate. Frank Stamford speaks of the astonishingly rapid rate of change in forest life: 'The whole nature of what they were doing there was so temporary, it is inevitable it would get lost'. He is annoyed by collectors who take things that will last, such as dogspikes, rail joiners, saw blades, remains of beds, crockery, bits of galvanised iron. He believes that the tramway formations and rights of way are worth preserving, but, he continues, 'I almost think it's immoral if one tries to preserve something that's going to fall apart in the next 50 years anyway … if it prevents industry from going on. After all, what were all those tramways there for?'[15]

Preservation seems artificial in the bush. The forest, although not necessarily the same forest, does recover, and quickly. The regenerative powers of forests enter the arguments over future forest utilisation. If the forests swallow the marks of intruders so easily, if they renew so rapidly, then history seems to serve the developer. 'Few traces of these tramways – once such a feature of the mountain forests – are now visible', wrote A W Shillinglaw of the Forests Commission of Victoria in 1959. He was presenting a submission on 'Forests utilization' to the State Development Committee, which was considering the advisability of logging water catchment forests. Shillinglaw commented on 'the remarkable resilience of Nature, which in the forested catchments, has healed the scars of Man's abuse over nearly a century of occupation'. Temporary disturbance of catchments, he pointed out, was not the same as permanent damage.[16]

A 1988 archaeological study of a sawmill site in the Erica district drew similar conclusions: 'In only 50 years the entire area investigated has returned to natural forest cover, with the exception of the sawdust heap'.[17] These evaluations are purely aesthetic and give value to any sort of vegetative cover, but even botanist David

Timber was cut at the New Federal Mill between Powelltown and Warburton from 1935 to 1949 and the last of the huts collapsed at the end of the century. (Photo: Tom Griffiths)

Ashton says 'You'd be very hard put to go back to old logging sites now, 40 to 50 years old, and say that this wasn't natural'.[18]

What is natural and what is cultural has become a distinction charged with significance in modern debates about the environment. The 1990 Land Conservation Council investigation into wilderness in Victoria valued land according to its remoteness and naturalness; it consequently underestimated Aboriginal impact on the environment and regarded historic features as negatives in 'aesthetic naturalness'.[19] Developers, on the other hand, seize on the history of utilisation to justify future exploitation, as if the discovery of Aboriginal environmental manipulation frees future inhabitants from any further obligation to the land.

A timber industry submission to the Helsham inquiry into Tasmania's Lemonthyme and Southern forests in 1987 argued that forests constituted 'cultural heritage' because roading, mining and logging had taken place 'for a considerable time'. And anyway, for thousands of years before that Aborigines had hunted and fired the forest. Logging and reforestation, it was dubiously submitted, would continue this tradition of cultural intervention. A 'natural' forest was artificial.[20] In such debates, history becomes only a political tool, rampantly relative and unattached to locale.

Modern miners also draw on history to justify their work. They are probably the first of their kind to find their prospecting activities complicated by growing community respect for the relics of their precursors. In the mountains, mining historians are miners first and historians second. History is the best prospecting tool they have. More effective than a pick or even a metal detector are the nineteenth-century quarterly reports of the government Mining Surveyor. Grubbied photocopies of these documents are pinned to the walls of mining site offices. And what better respect to pay to the past than to refurbish an old battery and move it from an abandoned mine to a modern active one? After all, the miners point out, the machinery has a history of mobility. History is a help, but heritage a hindrance.

Despite some appeals to history, industry generally remains shy of it. Historic features have been sacrificed in the timber industry's recent attempts to smarten up its public image. The last link with the steam era was demolished at the Powelltown mill in 1988. Three boilers installed at the very beginning of the town in 1912 to supply steam to the mill were dismantled, and their foundations demolished. They were found by a landscape consultant to be an 'eyesore'. When history is not ugly, it is dangerous. Many tramway bridges were blown up in the 1960s, 'to make them safe' and to provide the army with explosives practice. More recently a historic mountain hut was under threat of demolition because park managers worried that it might lead bushwalkers away from safe hiking paths.[21]

History was slow to find its way into land management considerations. It first did so through the identification of discrete sites and buildings, but historians were rarely involved in this work. The earliest historical reports for Victoria's Land Conservation Council in the 1970s were generally prepared by architects. They showed a preference for built structures and emphasised visual over documentary criteria of significance.[22] Archaeologists, like architects, brought a technical expertise to the analysis of surviving material culture and confirmed the secondary status of 'indoor' research.

The first cultural resource surveys in United States forests in the late 1970s involved little consideration of history. 'All researchers carried shovels or trowels', recalled historian Theodore Karamanski. 'If it rained, we all ran indoors and the principal investigator would turn to me and say, "Today you can do historical research."' It was not long, however, before weather came to play a less determining role. Research by shovel alone was gradually supported by documentary and interview work. Historical research, it was found, could help locate potential sites of settlement and industry in the forests and could also provide a context for assessing the significance of such sites.[23]

However, history need not be confined to the interpretation of circumscribed places with observable relics. It has a more fundamental role in an environment

EXHIBITING THE TALL FORESTS

LUKE SIMPKIN

Melbourne Museum is a large museum complex which opened in 2000 opposite the nineteenth century Royal Exhibition Building in Carlton Gardens, Melbourne. The striking modernist design accommodates an institution with 150 years of history in researching, collecting and exhibiting Australian nature and culture. The surprising permanent centrepiece of the new museum is the Forest Gallery – a living outdoor space focusing on the tall mountain ash forests of Victoria. Melburnians have long had a fascination with this distinctive, impressive environment, and there is even a history of exhibiting these forests on this site.

So why put a forest in a museum?

We didn't.

Like the real forest, it has trickling water, steep fern-clad slopes, towering trunks, tiny flittering birds and the smell of damp humus. *Unlike* the real forest, it has underwater viewing into a stream, a striking audiovisual display of bushfire, an antarctic plant fossil next to its living cousin and a representation of the seven Kulin seasons of the Yarra Valley. The real forest is far larger and more complex, of course, but its secrets are not always evident to the casual bushwalker. Through this exhibition, the museum aims to provide a perspective on the tall forests that it is impossible to get by visiting them yourself. However, when you do next visit the real thing, you will see them differently. And when forest management next arises as an issue, you may take a stronger interest. It is a 'shop window' on the tall forests.

It is also a 'shop window' on the museum. Museums are commonly defined by what they *have* rather than what they *do*. Far from the archetypal white-coat-clad dusters of stuffed animals, museum staff actively research wildlife populations, evolutionary histories, cultural traditions and technological change. Museums monitor and interpret change, and it is change in the living world that forms the focus of the *Forest Secrets* exhibition.

The gallery is divided into five zones, each exploring one of the five major agents that precipitate change in the tall forests:

- *Water* explores the role and effect of water as it shaped the landforms and created the conditions for forest life.
- *Earth* explores the deep history of continental drift and the resulting patterns of evolution of life.

Lush vegetation contrasts with the museum's modern architecture. (Museum Victoria)

- *Climate* explores the seasonal responses of living things to climatic cycles.
- *Fire* explores the essential force of bushfire as both cataclysm and the genesis of new life.
- *Humans* explores how cultural perceptions of the forest have shaped the interactions between people and nature in this region.

This unusual exhibition brings together a multimedia exhibit, a live animal display, sculptural artworks and a botanic landscape. The established trees and shrubs, some species of which have never been transplanted before, were trucked from a forest road construction site, logging sites, and a tree farm to the museum. A further 6000 indigenous plants were propagated for the gallery landscape.

Forest Secrets is a new kind of museum exhibit. In blending some of the qualities of zoos and botanic gardens with traditional museum expertise, it seeks to create a challenging and exciting experience for visitors, and a lasting 'window' into the magnificent tall forest environment.

where trees live longer than their managers, where natural and imposed forest cycles are greater than a lifetime. Foresters work within a management timescale that demands a history; yet they jealously sequester their sources. They have turned to history to defend themselves against the attacks of environmentalists; they have also frustrated outsiders inquiring into patterns of forest utilisation. 'Bureaucratic secrecy' has led to the suppression and loss of historical evidence.[24] Foresters and their domain are now under scrutiny, by historians as much as by 'greenies'. In 1988 the Australian Forest History Society was formed.[25]

And what of the material culture of the forest itself, the physical archive of the biota, the sap of history? Ecologists sift the forest biomass for clues to the past as if they were archaeologists. They are looking under the ferns and beneath the bark for another sort of heritage, the historical forest. Because of the significance of Black Friday in making and shaping the forest today, scientists are increasingly devoting their energies to understanding this defining fire. Stretton's job was to find out why it burnt. Ecologists are interested in *what* it burnt. What was the forest like before 1939? In what ways was it already greatly modified by a half-century of intensive sawmilling and a newly introduced fire regime? How did the two decades of salvage logging that followed it further corrupt the living biological record?

As we have seen, David Lindenmayer is one of those who is systematically scrutinising 'the biological legacies' of that era. He is fascinated by the architectural heritage of the forest. He has thousands of survey plots in hundreds of sites in the Central Highlands in order to record variations in stand structure, numbers of fire scars, and the stem diameters of fire-killed trees. This is a vast statistical time machine, reading back to 1939. 'The problems', he says, 'can be solved through careful management of the architecture of the forest'.

History is clearly up for debate in the evaluation and management of predominantly natural areas. Historians can help not only by identifying places and sites of cultural significance, and not only by messing with the concept of 'natural'. A historical perspective can serve a powerful integrative function in environmental studies. History – as much as ecology – is the key to enlightened management of the forests of ash, as Stretton knew. And the biological heritage – green and fibrous – is a historical document as significant as the pale and fibrous pages of testimony presented to his Royal Commission.

EPILOGUE

'IT'S BLACK FRIDAY' screamed the *Sydney Morning Herald* in large bold type on 7 January 1994 as fires burned along the north and south coasts of New South Wales and even invaded the harbour city. It *was* a Friday, and hundreds of thousands of hectares of forest were indeed blackened, but it was an event 55 years earlier, and in another state, that the phrase evoked. The *Sydney Morning Herald* made a grave and provocative comparison.

Fires are strangely historical. They inspire stories, disturb dreams and evoke memories. They discover ordinary heroes and flush out extraordinary villains. They inflame blame. They devastate and punctuate lives, they invite comparison and they are named. The New South Wales fires, too, made their bid for a place in history. They claimed the horrific stature of their own 'Black Friday'. In the wake of the Gulf War, they were 'the mother of all fires'. They were the worst in 50 years, they drew firefighters and equipment from every Australian state and territory, and they brought upon NSW as a whole its first ever total fire ban. They were more extensive but less intensive than Ash Wednesday, 1983; they engulfed almost as many hectares as Black Friday, 1939; they spared human property and life much more than either. And they fanned the flames of all the recurrent debates in Australian society about fire and our responsibilities to it and for it.

The New South Wales fires were mostly deliberately lit. 'The hand of man' was again at work. Was it arson or ignorance? Picnickers barbecued their sausages at the Royal National Park south of Sydney despite the total fire ban, remaining even as wildfire bore down upon them. The discovery of human implication in 1939 was a shocking finding then, just as it was in New South Wales in 1994. We have an image of vindictive, secretive evil-doers skulking about the bush and reeking of petrol and malice aforethought. But in 1939, Stretton condemned not malevolence but indifference, not criminality but endemic pyromania.

Sydney's experience of fire prompted condemnation of those who lit fires deliberately and carelessly, and also of those who did not light them often enough. Within days of the first major fires, the New South Wales Farmers Federation and a National Party member of parliament had flushed out a new villain. They claimed that firefighting had been 'severely hampered' by closed-off fire trails and 'almost non-existent' fuel reduction burning in national parks. This was said to be due to the influence of 'greenies', whose advocacy of aesthetic values, species diversity and ecological integrity had allegedly curbed control burning in forests and parks. Mr Phil Koperberg, Commissioner of NSW Bush Fire Services and one of the generation who had pioneered control burning, believed that only about 10 per cent of the area that used to be burnt off in the mid-1970s had been 'hazard-reduced' in the 1990s. 'The environment has lived with fire for 40 000 years', he declared, 'and we come along in 200 and think we know how it's done, but we don't.'[1]

The change had come partly because of the pressure of community complaints about air pollution, from people bridling against their acrid inheritance of a 'continent of smoke'. But, as we have seen, control burning had also undergone its own revolution since 1970. Foresters began to accept that the sort of burning best suited to the protection of timber and humans might not also be the best for plants and animals. In New South Wales it was widely accepted that fuel reduction burning had been overdone in the 1970s. And it was misleading to suggest that the green movement trenchantly opposed 'the red steer'. Some ecologies depend on periodic holocaust fire, and 'control burning' could be too controlled. Fire – or the right kind of fire – encouraged biological diversity. And the right kind of fire for a particular environment depended (among other things) on its intensity, frequency and seasonality. National and state-wide arguments about fire founder hopelessly on the realities of its intensely local history and ecology.

But there is little room for such niceties in the heat of the moment, particularly when human life and property are threatened. Military metaphors abound, and the enemy appears monolithic, wilful and contemptuous of particular ecologies.

Sydneysiders of 1994 saw the fire-front as the 'greatest threat to national security'. The firefighters were hailed as our modern Anzacs. Those who allegedly put them in danger by encouraging the enemy – by stamping out control burning – were traitors, as bad as the arsonists.

There will be more Black Fridays, and they're not all black. We have to accept them and plan for them, like drought. We should aim to survive them, even if we can't hope to prevent or control them. On that criterion, the New South Wales fires were mostly a triumph. The loss of human life and property was miraculously small considering the size and scope of the fires. And although they prompted yet another national debate about greenies, fire and control burning, the fires also flushed out some distinctly local politics. The green movement was threatening on another front in New South Wales early in 1994, and there was political mileage in smothering it. As flames licked along the coast towards Sydney, the state government was considering controversial plans to create new wilderness areas.

■

The original Black Friday was truly extraordinary. It was an extreme fire in a dangerous forest in the 'great fire flume' of south-eastern Australia.

In 1939 Stretton found that the *experience* of the past could not save those who dwelt in the forests of ash. Their own human lives were too short, dwarfed by the towering biographies of the trees. To survive in this environment they needed meanings, stories and strategies bigger than themselves. So Stretton conducted a social and ecological enquiry in the hope of transforming experience into history. 'What might, and did, happen' could only be understood through a cross-examination of culture and biology in the singular domain of this parochial timber.

Throughout his career, Judge Stretton was concerned with what constituted a moral community, a robust civic culture. He practised for ten years as a solicitor, went to the Bar in 1929 and, in 1937 at the age of 43, became the youngest judge appointed to the County Court of Victoria. He was Chairman of General Sessions and the senior judge and, in 1951, an acting Justice of the Supreme Court of Victoria. He declined permanent appointment to the Supreme Court, preferring his impressive range of civil duties. From 1938 he was foundation member and then chairman of the Workers' Compensation Board in Victoria and drafted the influential *Workers'*

Compensation Act of 1946 and the consolidating Act of 1951. He was also a judge
of the Court of Marine Inquiry and president of the Industrial Appeals Court.[2]

One of his other duties as a Royal Commissioner was to inquire into the causes
of fires that had threatened the Victorian State Electricity Commission (SEC)
coalmining township of Yallourn in Gippsland in 1944. In February of that year,
13 people died in surrounding areas, hedges caught alight in town, patients were
evacuated from the hospital, and trucks at the railway station filled with briquettes
started to burn fiercely. Embers from the fires also landed on the exposed coal in
the open-cut near the town, and fires broke out simultaneously over a wide area
of the mine and continued to burn for several days.[3]

Stretton believed that the protection of a town such as Yallourn – a critical site of
power production, especially in war-time – depended on the goodwill and willingness
of its people to respond in a time of crisis. He found instead an impoverished civic
culture which he attributed to the 'suffocating paternalism' of the SEC's adminis-
tration of the town. 'Here indeed', Stretton thundered memorably, 'the townsman
enjoys all that the heart of man may desire – except freedom, fresh air and independence'.[4] The lack of municipal democracy and home ownership produced a
'reluctance to engage in common endeavour' and discouraged a sense of social
responsibility. As a result of Stretton's report – ostensibly about fire but ultimately
about moral community – townspeople found a new political voice, and elections for
the first Yallourn Town Advisory Council were held in December 1947.[5]

In his 1946 Royal Commission into Forest Grazing, Stretton coined a conservation slogan when he preached of 'an inseparable trinity – Forest, Soil, and Water'.
He urged the cultivation of a 'forest conscience'.[6] Of environmental offenders, he
wrote: 'The man who by his carelessness or wantonness disturbs the delicate natural
balance between forest, soil and water works on a scale of wrong-doing which by
comparison makes the pickpocket a mere amateur fumbler.' But he concluded that a
'strongly, actively expressed public sentiment will do more to discourage the destroyer
than will fine or other punishment.'[7] In court, Stretton was severe on crimes of
violence but otherwise advocated leniency for young first offenders.[8] In 1952 he
was rebuked by the Victorian cabinet for his scathing comments on conditions at a
Watsonia emergency housing camp which he believed tended to produce criminality
among residents; he refused to convict one whom he considered 'more sinned against
than sinning'. The camp was closed soon afterwards.[9]

His 1939 inquiry into Black Friday found wickedness and wantonness, and demanded watchfulness and punishment. The hand of man was everywhere. 'These people have to burn the scrub to live', explained James Francis Ezard, and he did not just mean to make a living. He meant to survive. Otherwise 'they would be wiped out'.[10] There was continual discomfort at the Royal Commission and it wasn't just the extreme heat. As historian Stephen Pyne put it: 'It was as if Australians could not bring themselves to admit the reality of what had occurred or to confess the full complexity of the tragedy or their collective complicity in it.'[11] Stretton certainly acknowledged that 'the truth was hard to find'.

Discovering the causes of such fires was like analysing the first seconds of a stockmarket crash. Everyone was implicated, but who acted first? And did it matter? The panic was real and it was ready to precipitate. Fires were lit and fires were not put out because many people in the bush, frustrated by the rules of the Forests Commission, felt that it was better to burn late than never. Stretton called it mass suicide.

When, years later, the eminent historian Sir Keith Hancock wrote a book regarded as a pioneering environmental history of an Australian region, *Discovering Monaro* (1972), his model for such work was not a fellow academic but a judge, Leonard Stretton. Like Stretton, Hancock called 'witnesses' to bear testimony about the land, and offered an extended historical and philosophical parable about 'man and nature'. Both men projected powerful moral visions informed not only by environmental politics but also by the Bible. Hancock hoped that his own professional history would have some of the practical wisdom, bureaucratic fearlessness and commitment to applied ecology that he admired in Stretton's work as a Royal Commissioner.[12]

■

'Black Friday' became a portable metaphor for awesome tragedy, a standard for disaster, a code for nature engulfing culture. It is hard to write fire history without being melodramatic, without depicting human powerlessness in the face of the elements, without indulging the passive voice and portraying simple oppositions, a primal morality. My first draft of the history of these forests was a story of settlement, it was a story of what humans did to the forest. It told of how people had perceived, lived in and used the forests. It offered a history of farming, mining, sawmilling, tourism, and water and forest conservation. It recounted a human history of fire. It imagined all the drama and horror of Black Friday. It recognised Black Friday as a European creation; an awful consequence of a century of white settlement and environmental practice. It echoed the finding of Stretton's Royal Commission and elucidated the hand of man.

This was partly because my history had its origins as a historic site survey for Victoria's Land Conservation Council. It was our brief to separate nature and culture; natural scientists were already surveying the biological and geological values of the forests of ash, and so it was indeed our job to discern, by contrast, 'the hand of man'. At first, then, the forest remained the backdrop to my human story, a picturesque setting, a valued resource, something that was exploited, used, protected and acted upon, but which was rarely allowed a dynamic of its own. That is the way in which historians usually include the natural world in their narratives, by making it into an artefact, by drawing it within the sphere of human influence and diminishing its natural dimensions. Hence we often talk of 'impact' or 'land-use' or 'conquest'; words that describe a one-way relationship.

But then I began to look at the forest itself more closely. I began to realise that it was a community of trees. The trees had names and life histories, too. And fire in this forest was different to fire in other forests. Why? Was Black Friday entirely a European creation, or was it part of an ancient cycle essential to the maintenance of this natural community? How could I know unless I extended my enquiry backwards and sideways, backwards through time and sideways across disciplines? I could not write this forest's history without understanding, at least in a rudimentary way, its nature. I began reading about fire ecology, and I started talking to ecologists.

David Ashton's research into the private lives of the trees revealed that mountain ash forests perversely need a catastrophe to survive. They need Black Fridays. But they need them at long intervals, every few hundred years. Black Friday was, then, *not* entirely a European creation; it was part of an ancient natural cycle essential to this community of trees. Investigating the nature of ash enabled me to go beyond a simple appreciation of them as tall trees to a realisation that their ecology was the key to many of the region's cultural patterns – its dramatic fire history, the chronology and sequence of forest utilisation, the placement of bush sawmills, the professional anxieties of foresters, the imperatives of water supply managers. Ecology is not entirely a separate realm of specialist study; it is a systematic distillation of the sort of knowledge or bush lore that anyone who lives or works in a forest has to have. So ecology helped me plumb not just natural systems but cultural understandings.

But, as I journeyed into the ecology of Black Friday, history asserted itself again. The fires of 1939, although part of a long-term natural cycle, *did* constitute a unique event in significant ways. We are only beginning to understand just how unique, just how historical. Black Friday was an uncharacteristically extensive wildfire because it was indeed lit by the hand of man. It has changed the structure of the forests of ash for hundreds of years. It burnt more than three times the combined area of forest affected by any other fires in Victoria during the past century, and an area more than ten times larger than the annual average.

The historical shadow and ecological legacy of Black Friday, compounded by the effects of almost two decades of salvage logging, have been so dominating that we cannot be naturalists without being historians too. We may have underestimated the earlier prevalence of low intensity, ground-surface fires. We have probably given insufficient recognition to the different fire regimes that once produced more multi-aged ash forests than exist now. So Black Friday continually emerges as an intriguing artefact of nature and history, a cultural exaggeration of a natural rhythm. Even as we discover its ecological depth, we are reminded of its historical specificity.[13]

.

The writing of Australian history has always been suffused with a sense of the land and its difference – the observed peculiarities of antipodean nature – but it was not until the 1970s and '80s that 'environmental history' as a conscious sub-discipline emerged. Its sources are many: the imperial encounter with rapid environmental change in the New World, an emphasis by colonial and early national historians on the challenges and politics of land settlement, a long-term investment in the study of Australian land use by historical geographers, the geographic inspiration of the French Annales historians (especially Fernand Braudel), the post-war discovery of Australian antiquity, the rise of green politics since the 1960s, and the recent disciplinary definition of 'environmental history' in American scholarship.[14]

'Environmental history' is a term that is often used quite differently by scientists and humanities scholars. In one sense, geologists have been doing environmental history since Hutton and Lyell; biologists have been doing it since Darwin and physicists since Einstein. The discovery of an ancient age for the earth – and for humans – demanded a history. It became a habit of the nineteenth century to compile chronologies, make narratives, and construct genealogies. And so the environmental sciences strengthened as observational, experiential and storytelling disciplines. Charles Darwin was in accord with his society to the extent that he was engaged in a search for origins, a genealogical enterprise of immense and threatening proportions. A historical view of nature emerged, and contingency, uncertainty and history became embedded (albeit uneasily) in the scientific method.[15]

Until the early 1970s, 'environmental history' was a term often used by geologists, palaeobotanists and archaeologists in their analyses of environmental change in the quaternary period. In this context it was a phrase only incidentally applicable to human experience. But for some scholars, the economic life of early humans was indeed most fruitfully considered in relation to the wider economy of nature.[16] And for about a century, it was historical and cultural geographers who carried the burden of analysing *Man's Role in Changing the Face of the Earth*, as one landmark book put it in 1956.[17] In a perceptive and wide-ranging analysis of the origins of environmental history, Richard Grove, the author of *Green Imperialism* (1995), characterised the new environmental history in its initial stages as 'a fairly parochial takeover bid by North American scholars of what was already a firmly established subject'. Grove enjoyed making fun of the innocence of American environmental historians such as Roderick Nash who, in 1972, behaved 'as if discovering something quite new' and lamented a lack of reading material in the area.[18]

Historians have indeed been slow to integrate an active nature into their narratives and to acknowledge their interdisciplinary heritage in this field, but it is also true that many geographers were retreating from environmental history about the time that historians were venturing into it.[19] In any case, there are some distinctive dimensions to the latest generation of environmental history. For a start, it is more self-consciously *history*. It gets its sense of innovation from its explicit intervention in a discipline that is tenaciously human-centred. Environmental issues have generally been assigned to the other side of the great divide between the sciences and the humanities, and historians – by inclination and training – tend to feel more at home with books and paintings than they do with pollen and pipettes. The former are therefore more likely to enter their historical analyses as evidence than the latter. When history has aspired to be a science (a periodic goal), it has been in method rather than subject matter, whereas geography has had to accommodate a constant identity crisis in both matter and method. History has remained unambiguously a humanities subject, concerned overwhelmingly with human morals and motivations, a professional discipline formed in the nineteenth century at the time of rising nationalism and increasing bureaucracy, and therefore especially concerned with identity and documents.[20] As Alfred Crosby has noted, historians were trained to value written eye-witness accounts above all else, and yet environmental history was often unobserved and unrecorded.[21]

But that is not a new challenge for historians, whose discipline – literally – is to read against the grain of their sources, listen for silences, and speak for the forgotten and oppressed as well as the articulate and powerful. Now they are determined to give the non-human world some agency in the historical narrative.[22] So one of the ways in which the latest environmental historians have explained the emergence of this 'new' field has been by describing an expanding circle of empathy and ethics, one that now goes 'beyond the human dimension to embrace all life'.[23] 'This would indeed be history "from the bottom up"', enthused liberal progressivist Roderick Nash in 1972, fitting it into the framework of New Left history, 'except that here the exploited element would be the biota and the land itself'.[24] And the biota was no mere passive servant – no more than those other neglected agents being written into history: the working classes, blacks, women.

The revisionist histories giving voice to the experiences of these groups recognised the strength and subversity of their underworlds. So too, then, did environmental historians seek to describe a natural world that had rhythms and histories of its own, wanting to recognise nature as an actor in its own right, as more than a static physical base, more than a cultural construct.[25] They made a bid for nature to join the hallowed trio of race, class and gender as recognised determinants of human consciousness and social behaviour.[26]

Environmental history, constantly wrestling with biological determinism, also represented a return of materialist styles of history following the decline of Marxist history and economic history.[27] And it was powerfully shaped, and given urgency, by green and black politics which both emerged strongly in the late 1960s. Thus the historians' traditional concerns of identity, agency, economy, politics and (in the American case especially) nationalism helped them make environmental history new, and yet recognisably their own.

As Donald Worster put it, environmental history was 'born of a moral purpose, with strong political commitments behind it'; it is a scholarly response to the sense of global ecological crisis.[28] American environmental historians reciting the history of their particular field of practice sometimes go back to George Perkins Marsh, the author of *Man and Nature* (1865), or early wilderness advocates Henry David Thoreau and John Muir, but more often they begin with the Wisconsin wildlife biologist and conservationist Aldo Leopold who, in his classic text *A Sand County Almanac* (1949), called for 'an ecological interpretation of history'. The next landmark in this lineage is generally Rachel Carson's *Silent Spring* (1962), a book celebrated as being 'both scientifically informed and evangelical', linking ecology and politics in an explosive way.[29] A combination of romanticism and critical politics, and a strong dose of apocalypticism, has given the new field its character.[30] The dropping of an atomic bomb on Hiroshima in 1945 and the landing on the moon in 1969 – and the sight of the Earth floating alone, finite and lovely in awesome space – dramatically enforced a planetary consciousness.[31]

Environmentalism and environmental history are, of course, different phenomena and they are often in tension with one another. When historians write of giving nature a voice, they seem to align themselves with environmentalism. And, indeed, one way that historians first ventured into this field was by writing about wilderness consciousness and conservation politics. Yet environmentalism often separates and opposes the categories of nature and culture which environmental historians are so eager to enmesh. And environmentalism gains a critical political edge by presenting itself as new, as 'green', whereas historians continually find earlier conservationists and more ancient origins of environmentalism.[32]

The new environmental history's explicit engagement with the scientific insights and metaphors of ecology is one of its most distinguishing features, and it commits the field to sharing an intriguing intellectual and political journey. Ecology took form as a discipline from the 1920s, and became powerful in popular culture from the 1960s. Libby Robin has analysed the rise of ecological consciousness in Australia in this period through a study of the Little Desert dispute in Victoria in 1969.[33] She shows how ecology gained confidence and definition as a science partly through a close engagement with post-war conservation politics, but that it became a looser and more volatile concept in the hands of the green movement of the early 1970s. A dramatic political and generational change overtook conservation in Australia at that time, and the uses of ecology were caught up in it. And ecology has wrought its own revolutions. When it burst upon the public imagination of the Western world in the mid-twentieth century, it appeared to be a study of equilibrium, harmony and order. At the end of the century, however, it had become more concerned with disturbance, disharmony and chaos.[34] Suddenly scientists were looking everywhere for disturbance in nature, especially signs of disturbance that were not caused by humans but by the restlessness of climate and the violence of the elements.[35]

When preaching 'the balance of nature', ecology had seemed a natural ally of conservation. But 'disturbance' seemed to justify intervention. As Worster put it, 'What does the phrase "environmental damage" mean in a world of so much natural chaos'?[36] Ironically the change came about at least partly because humanities scholars insisted that 'disturbance' and historical change could not be eliminated or wished away, they had to be studied. Stephen Pyne wryly observed that '[a]n environment without people is even more abstract and meaningless than the ideal frictionless surface beloved of physicists.'[37]

So environmental historians follow the fashions of ecology with special interest – from climaxes and superorganisms to energy flows and ecosystems to patch dynamics and landscape mosaics. They also speculate about the politics of an ecology of community and cooperation as against an ecology of individuality and competition.[38] It can be disturbing to find science as malleable as culture, but it can also be strangely satisfying too, especially to humanities scholars, weary of coping with the charges of relativism. Richard White wrote in 1990: 'Historians thought ecology was the rock upon which they could build environmental history; it turned out to be a swamp.'[39] And in 1998, Stephen Pyne reflected mischievously that '[e]cological science is far too unstable to serve as a foundation for history. Rather it furnishes convenient scaffoldings to be erected, torn down and reassembled on the hard pilings of philosophy and art. The rest will be swept away in the next storm of discovery and paradigm shifts.'[40] The forests of ash are a privileged theatre in which to observe the ecology of ecology.

Environmental history in its new guise is also distinctive in its strong identification with the humanities, especially in its commitment to narrative. The pioneering Australian historical geographer (and environmental historian!), J M Powell, an astute commentator on tradition and innovation in the new field, has discerned 'a greater emphasis on story-telling'.[41] The commitment to narrative brings environmental historians back to their fundamental disciplinary concern with values and morals, and requires that they reflect upon what this new materialism and old determinism does to the human story. Having negotiated the new science of chaos, they return to their own side of the great divide and must deal with the linguistic turn in literary scholarship. Does the enrolment of an active nature in the narrative allow oneself to tell different kinds of stories? How do we 'animate nature without anthropomorphising it'?[42] If narration is central, then just how plural are its possibilities, and how do we judge multiple truths?

In a stimulating essay, William Cronon has wrestled with the implications of postmodern critical theory for environmental history, an important task in a field where historians often make a theatre of their realism, are invigorated by the bracing otherness of nature, relish their muscular engagement with the physical archive, enjoy putting on boots to go to work, and often find themselves involved in practical, political outcomes from their research and writing. He concludes that we must tell not just stories about nature, but 'stories about stories about nature ... because narratives remain our chief moral compass', and because the stories we tell change the way we act in the world.[43] Scholars in the humanities, especially when they are working in an interdisciplinary setting, also need to advocate the distinctive skills of the storyteller, to defend the logic of poetry, and to champion narrative not just as a means, but a method, and a rigorous and demanding one.[44]

Environmental history, then, builds on (and does not displace) the contributions of disciplines that have long invested in the same subject, such as archaeology, geography, forestry, ecology and economic history.[45] But it is also, in certain ways, a distinctive endeavour, recognisably a product of the late twentieth century. Its academic vitality springs directly from a contemporary sense of crisis about the human ecological predicament. It often moves audaciously across time and space and species and thereby challenges some of the conventions of history by questioning the anthropocentric, nationalistic and documentary biases of the craft. Environmental history frequently makes more sense on a regional or global scale than it does on a national one. It uniquely bridges planetary and deeply local perspectives, staking a claim for histories that are bound intimately to place and also embrace the natural world; histories that are deeply attentive to human and biological parochialism. Environmental history offers a genuine and challenging meeting of sciences and the humanities. It is a place where social history and deep time have to find their correspondences. Yet environmental history remains, at heart, one of the humanities, concerned with cultural, moral, economic and political questions, and founded in narrative.

So this book consciously weaves between the living forest and the paper trail, the trees and the words, telling stories that enable us to act, and exploring the morality and politics of fire, as well as its ecology. It is no accident that its main characters – alongside the measured and titled giants, the named fires, the Aboriginal residents, and the miners, sawmillers, foresters, bureaucrats and naturalists – are a judge and an ecologist. Stretton and Ashton engaged with fire and ash, and made this forest luminous with morals and meaning. This is one of the most intensively studied bio-regions in Australia, and Stretton's commission and Ashton's plots, both seeded by Black Friday, constitute possibly the most valuable archives of environmental history in this country. The fate of the forest depends upon them.

CONTRIBUTORS

LINDY ALLEN, Senior Curator, North Australia in the Indigenous Cultures Department at Museum Victoria, has worked for more than 20 years with Aboriginal communities in the Northern Territory, Queensland and Victoria.

BILL BIRCH is Senior Curator in Geology. The origin of Victorian gemstones is one of his major research interests.

JOAN DIXON, Senior Curator of Mammals, has published widely on Australian mammals and their history, in her 36 years as a member of museum staff.

ROSS FIELD is an entomologist who worked at Museum Victoria as the Director of Natural Sciences and Director of the Environment Program. He was responsible for the development of the Forest Gallery at the museum, a living representation of the mountain ash forests to the east of Melbourne.

Museum Victoria's first Curator of Fishes, MARTIN F GOMON has helped build the museum's fish collection to the point where it is now Australia's third largest. Although responsible for freshwater and marine species, his research focus has been directed primarily toward the taxonomy and biogeography of a number of marine families.

RORY O'BRIEN is Assistant Collection Manager in the Ornithology Section. His interests in Ornithology are diverse, and he has contributed to studies on plumages, moult, zoogeography, island biogeography, seabird biology and historical ornithology.

GARY PRESLAND was formerly Head Curator of the Technology Program at Museum Victoria. He is currently the 2000 Thomas Ramsay Science and Humanities Fellow.

LUKE SIMPKIN, formerly producer and now manager of the Forest Gallery and Live Exhibits, has long had an interest in environmental education and Victoria's tall forests in particular.

KEN WALKER is Senior Curator of Entomology at Museum Victoria. He has been with the museum for approximately 20 years and conducts research on the native Australian bee fauna and the plants they pollinate.

ELIZABETH WILLIS is a senior curator in the Australian Society and Technology Department at Museum Victoria. She enjoys bushwalking, exploring country towns, and researching visual representations of Australian identity.

ALAN YEN is a zoologist with research experience on insect–plant interactions and on the ecology of terrestrial invertebrates. He graduated from La Trobe University and has worked at Museum Victoria with the Invertebrate Survey Section.

ACKNOWLEDGEMENTS

This is a substantially revised and enlarged version of my earlier book, *Secrets of the Forest* (Allen & Unwin, Sydney, 1992). That book stemmed from a fruitful collaboration between what was then the Historic Places Branch of the Victorian Department of Conservation and Environment and Monash University's Master of Arts in Public History Program. I would like to extend my thanks again to all the people acknowledged in *Secrets of the Forest*, especially those whose essays about historic sites formed Part II of that original study. In particular, I again gratefully acknowledge Jane Lennon, Ray Supple, Graham Perham, Chris Smith, Anita Brady, Paul Barker, Peter Evans, Meredith Fletcher, Graeme Davison and Mark Tredinnick for helping me to find my way to, and through, the forest.

This new book was prompted by the interest shown by Museum Victoria staff in *Secrets of the Forest*, especially as they worked towards the opening of the Forest Gallery at the heart of Melbourne Museum. We agreed that the new Gallery, which features a tall, wet forest, presented a fine opportunity to revise, update and extend the original book. I feel very fortunate to have been able to do so, and I want to thank Museum Victoria staff, especially Luke Simpkin, Elizabeth Willis and Ross Field, for their enthusiasm and encouragement. I am also grateful to Penny Mules and Daniel Catrice for assisting us with picture research. Every chapter of the earlier book has been revised, and a number have been substantially extended. In addition, *Forests of*

Ash has three new chapters, a preface and an epilogue (constituting about 25 000 new words), plus twelve 'spotlights' written by Museum staff which provide windows on their research. I thank the contributors for being a part of this project and for sharing their knowledge so generously.

The research and writing of *Forests of Ash* has been completed while I have been in the History Program of the Research School of Social Sciences at the Australian National University. I want to thank my colleagues and students in the History Program for their constant support, and especially Greg Bowen for his able assistance with research.

I am very grateful to a number of people who have read all or part of the new manuscript, in particular Lindy Allen, Bain Attwood, David Bowman, Judy Clark, John Dargavel, Ross Field, Ray Griffiths, Kay Griffiths, Grace Karskens, David Lindenmayer, Joe Powell, Penny van Oosterzee, Libby Robin, George Seddon, Luke Simpkin and Barry Smith. Phillipa McGuinness secured this book for Cambridge University Press and it was a delight to work with her again. Peter Debus and Paul Watt ably guided it through production.

Libby Robin not only read all the words, but talked about them with me and made them better. Kate Griffiths and Billy Griffiths offer me wonderful friendship and encouragement. My parents, Ray and Kay Griffiths, first took me into these forests and we are still learning about them together.

One cannot write about mountain ash without thanking David Ashton. One cannot write about fire anywhere in the world without admiring Stephen Pyne. And one cannot write about Black Friday without blessing Leonard Stretton – and his daughter, Althea Williams, and one of his sons, Hugh Stretton, for talking to me about their father.

The forests of ash themselves remain an inspiration.

NOTES

PREFACE

1 L E B Stretton, *Report of the Royal Commission into the Causes of and Measures Taken to Prevent the Bushfires of Jan. 1939*, Government Printer, Melbourne, 1939.

2 The quote is from Stretton's report.

3 Stephen J Pyne, *Burning Bush: A Fire History of Australia*, Allen & Unwin, Sydney, 1992, p. 309. '[A] country that considered itself civilised' is also Pyne's phrase, p. 312.

1 CONTINENT OF FIRE

1 Mary E White, *The Greening of Gondwana*, Reed Books, Sydney, 1986, p. 43.

2 Alfred Russel Wallace, *Australasia*, 4th edition, 1884, quoted in Rod Ritchie, *Seeing the Rainforests in 19th-century Australia*, Rainforest Publishing, Sydney, 1989, p. 15.

3 Charles F Laseron, *The Face of Australia*, Angus and Robertson, Sydney, 1953, p. 19.

4 Paul Adam, *Australian Rainforests*, Oxford University Press, Oxford, 1994, p. v; George Seddon, 'Eurocentrism and Australian science', in his *Landprints: Reflections on Place and Landscape*, Cambridge University Press, Cambridge, 1997, pp. 74–5.

5 D M J S Bowman, *Australian Rainforests: Islands of Green in a Land of Fire*, Cambridge University Press, Cambridge, 2000, p. 41.

6 Quoted in Adam, *Australian Rainforests*, p. 137.

7 Introduction to volume 1 of *Flora of Australia*, 1981; E M Truswell, 'Australian rainforests: The 100 million year record', in L J Webb and J Kikkawa (eds), *Australian Tropical Rainforests: Science – Values – Meaning*, CSIRO Australia, Melbourne, 1990, pp. 7–22; Adam, *Australian Rainforests*, p. v; and Seddon, *Landprints*, pp. 74–5.

8 Penny van Oosterzee, 'Redefining notions of boundaries – philosophical, historical and scientific', in I Walters *et al*, *Unlocking Museums*, Proceedings of the 4th National Conference of Museums Australia Inc,

Museums Australia NT Branch, Darwin, 1997, p. 24.

9 Quoted in Hal Hellman, *Great Feuds in Science*, John Wiley & Sons Inc., New York, 1998, p. 146.

10 Quoted in Hellman, p. 153.

11 Richard Fortey, *Life: An Unauthorised Biography*, Harper Collins, London, 1997, pp. 217–20.

12 Hellman, p. 155; John A Stewart, *Drifting Continents and Colliding Paradigms: Perspectives on the Geoscience Revolution*, Indiana University Press, Bloomington and Indianapolis, 1990.

13 This is the word used by White in *The Greening*, p. 40.

14 Fortey, p. 270.

15 Stephen J Pyne, *Burning Bush: A Fire History of Australia*, Allen & Unwin, Sydney, 1992, p. 4.

16 I am drawing on Pyne, White and Eric Rolls, 'The nature of Australia', in Tom Griffiths and Libby Robin (eds), *Ecology and Empire: Environmental History of Settler Societies*, Keele University Press, Edinburgh, 1997, pp. 35–45.

17 Timothy F Flannery, *The Future Eaters: An Ecological History of the Australasian Lands and People*, Reed Books, Sydney, 1994, chapter 6.

18 L J Webb and J G Tracey, 'Australian rainforests: Patterns and change', in Allen Keast (ed.) *Ecological Biogeography of Australia*, Dr W Junk bv Publishers, The Hague, 1981, p. 609.

19 For some early explorers' accounts of Aboriginal environmental management, see Major Thomas Mitchell quoted in Rhys Jones, 'Fire-stick farming', *Australian Natural History*, vol. 16, 1969, p. 225, and Edward Curr quoted in G Blainey, *Triumph of the Nomads*, rev. edn, Macmillan, Sydney, 1983, pp. 78–9.

20 Sylvia Hallam, *Fire and Hearth: A Study of Aboriginal Usage and European Usurpation*

in Southeastern Australia, Australian Institute of Aboriginal Studies, Canberra, 1979, p. vii.

21 Jones, pp. 224–8. For an alternative view, see D R Horton, 'The burning question: Aborigines, fire and Australian ecosystems', *Mankind*, vol. 13, 1982, pp. 237–51, and Horton, *The Pure State of Nature*, Allen & Unwin, Sydney, 2000.

22 Sandra Bowdler, 'Rainforest: Colonised or coloniser?', *Australian Archaeology*, no. 17, 1983; R Ritchie, *Seeing the Rainforests in Nineteenth Century Australia*, Rainforest Publishing, Sydney, 1989, p. 22.

23 Bowman, *Australian Rainforests*, p. 277.

24 Bowman, p. 249. See R S Hill's critique, 'Attempting to define the impossible', in *Australian Geographical Studies*, November, 2000.

25 I adapt this phrase from A H K Petrie, P H Jarrett and R T Patton, 'The vegetation on the Blacks' Spur region: A study in the ecology of some Australian mountain *Eucalyptus* forests I: The mature plant communities', *Journal of Ecology*, vol. 17, 1929, pp. 223–48.

26 Laseron, p. 23.

27 Bowman, p. 1.

28 D A Herbert (1932), quoted in Bowman, p. 1.

29 D Clode, 'Forests of Fire', Museum Victoria, 1998, manuscript kindly made available by the author.

30 Here I have drawn on William Cronon, *Changes in the Land: Indians, Colonists, and the Ecology of New England*, New York, Hill and Wang, 1983, pp. 10–12, and Donald Worster, *Nature's Economy: A History of Ecological Ideas*, Second edition, Cambridge University Press, 1994, chapter 17.

31 Ian Watson, *Fighting over the Forests*, Allen & Unwin, Sydney, 1990; P Gell and D Mercer (eds), *Victoria's Rainforests: Perspectives on Definition, Classification and Management*, Monash Publications in Geography No. 41, Melbourne, 1992.

32 L J Webb and J G Tracey, 'Australian rainforests: patterns and change', in A Keast (ed.) *Ecological Biogeography of Australia*, Dr W Junk, The Hague, 1981, pp. 605–94, quoted in Bowman, p. 25. See also a series of four articles on rainforest by George Seddon in *Landscape Australia*, nos 3 & 4, 1984 and nos 1 & 2, 1985. The final part of this series

(2/85, pp. 141–51) is a study of temperate rainforest. For a lucid historical analysis of definitions of sclerophylly, see Seddon, 'Xerophytes, xeromorphs and sclerophylls: the history of some concepts in ecology', *Biological Journal of the Linnean Society*, vol. 6, March 1974, pp. 65–87.

33 Bowman, p. 26.

34 J M Gilbert, 'Forest succession in the Florentine Valley, Tasmania', *Papers and Proceedings of the Royal Society of Tasmania*, vol. 93, 1959, pp. 129–51, quoted in David Ashton, 'Fire in tall open-forests (wet sclerophyll forests)', in *Fire and the Australian Biota*, eds A M Gill, R H Groves and I R Noble, Australian Academy of Science, Canberra, 1981, p. 342.

35 Bowman, pp. 66, 133.

36 K W Cremer, 'Eucalypts in rain forest', *Australian Forestry*, vol. 24, 1960, pp. 120–6, quoted in Bowman, p. 268.

2 TALL TREES

1 D H Ashton, 'Tall-open forest' in R H Groves, *Australian Vegetation*, Cambridge University Press, Cambridge, 1981.

2 In what follows about the occurrence and ecology of the mountain ash, I have drawn on the work of David Ashton, in particular his 'Studies on the autecology of *Eucalyptus regnans* F.v.M.', PhD thesis, School of Botany, University of Melbourne, 1956, and 'Fire in tall open-forests (wet sclerophyll forests)', in *Fire and the Australian Biota*, eds A M Gill, R H Groves and I R Noble, Australian Academy of Science, Canberra, 1981, pp. 339–66. See also A V Galbraith, *Mountain Ash: A General Treatise on Its Silviculture, Management and Utilisation*, Forests Commission of Victoria, Melbourne, 1937. Hec Ingram's description of mountain ash appeared in *Early Forest Utilisation*, Forest Recollections series, Institute of Foresters of Australia, Victorian Division, Mitcham, 1979, p. 6.

3 The 'gleaming boles' are described by Nettie Palmer in *The Dandenongs*, Melbourne, 1953, p. 15; the ships' masts by Captain John Freyer (1850) quoted in A P Winzenried, *The Hills of Home: A Bicentennial History of the Shire of Sherbrooke*, Shire of Sherbrooke,

Melbourne, 1988, p. 35; and the streamers of bark by John Walters in his 'Botanical tour of the Dandenong Ranges, May 6th 1853', published in the London *Gardeners' Chronicle*, 24 September 1853, pp. 613–14, and quoted in Helen Coulson, *Story of the Dandenongs, 1838–1958*, Cheshire, Melbourne, 1968, pp. 4–5.

4 Jenny Mills, 'Boranup forest: old-growth or regrowth?', in John Dargavel (ed.), *Australia's Ever-Changing Forests III*, Centre for Resource and Environmental Studies, Australian National University, Canberra, 1997, pp. 112–20.

5 Stephen J Pyne, *Burning Bush: A Fire History of Australia*, Henry Holt & Co, New York, 1991, p. 49.

6 George Seddon, Review of *Secrets of the Forest, Australian Historical Studies*, vol. 26, no. 102, April 1994, pp. 150–1.

7 Marianne North, *Recollections of a Happy Life* (edited by Mrs John Addington Symonds), 2 vols, Macmillan & Co., London, 1892, vol. 2, pp. 144–5.

8 Quoted in Bernard Mace, 'Mueller – Champion of Victoria's giant trees', *Victorian Naturalist*, vol. 113, no. 4, August 1996, p. 204.

9 Geoffrey Blainey (ed.) *Oceana: The Tempestuous Voyage of J A Froude, 1884 & 1885*, Methuen Haynes, Sydney, 1985 (first published 1886), pp. 62–7.

10 *The Giant Trees of Victoria*, folio of photos prompted by the 1888 competition, Victorian government, Melbourne, 1890; M Carver, 'Forestry in Victoria 1838–1919', vol. E of 5 vols, unpublished typescript, Department of Natural Resources and Environment Library, Melbourne, n.d.; George Cornthwaite, 'Tallest trees recalled', *Walkabout*, reprinted in Jim Tigner, *Recalling 100 Years: A Brief History of Thorpdale and District 1876–1976*, 1976, p. 61. Parts of this book, such as this paragraph, have been drawn from my report to the Land Conservation Council of Victoria, *Fire, Water, Timber and Gold: A History of the Melbourne East Study Area*, Historic Places Section, Department of Conservation, Forests and Lands, Melbourne, 1989. Several paragraphs of this report were reproduced without acknowledgment in F R Moulds, *The Dynamic Forest*, Lynedoch Publications, Melbourne, 1991, pp. 8, 10, 78–9.

11 Crosbie Morrison, 'Wild life' radio script, 12 January 1944, series 3, Crosbie Morrison Papers, State Library of Victoria; Galbraith, *Mountain Ash*, p.17.

12 For this quote and a summary of the debate about the age of eucalypts, see Tim Bonyhady, 'Primeval forests in Australia', in John Dargavel (ed.), *Australia's Ever-Changing Forests III*, Centre for Resource and Environmental Studies, Australian National University, Canberra, 1997, pp. 24–34, and Bonyhady, *The Colonial Earth*, Melbourne University Press, Melbourne, 2000, chapter 9.

13 A G Campbell, Notes on photographs of giant ash (1942), Ornithological collections, Museum Victoria.

14 Marcus Clarke, Introduction to Louis Buvelot's 'Waterpool near Coleraine' (1874), in Bernard Smith (ed.) *Documents on Art and Taste in Australia*, Oxford University Press, Melbourne, 1975, pp. 133–6.

15 N Caire, 'Notes on the giant trees of Victoria', *The Victorian Naturalist*, vol. 21, no. 9, January 1905, pp. 122–8.

16 Caire, p. 122.

17 Geoffrey Munro, 'Cumberland Scenic Reserve', in Tom Griffiths, *Secrets of the Forest*, Sydney, 1992, pp. 190–4.

18 Ashton, 'Studies on the autecology of *Eucalyptus regnans*', introduction.

19 Ashton, 'Fire in tall open-forests', p. 362.

20 A H Beetham, 'Aspects of forest practice in the regenerated areas of the upper Latrobe Valley', Diploma of Forestry thesis, 1950, Department of Natural Resources and Environment Library.

21 Ashton, 'Fire in tall open-forests'.

22 Hoddle is quoted in H S McComb, 'Surveying the Yarra Yarra River', *The Australian Surveyor*, vol. 7, no. 4, December 1938, p. 245.

23 Brough Smyth is quoted in Catherine Upcher, *Northern Diversion Sites, Archaeological Survey: Preliminary Report*, Department of Conservation and Environment, Melbourne, 1991, p. 14.

24 Hilary du Cros, *An Archaeological Survey of the Upper Yarra Valley and Dandenong Ranges*, Victoria Archaeological Survey, Melbourne, 1988, p. 34.

25 Daniel Bunce, *Travels with Dr Leichhardt* (Melbourne, 1859), reprinted by Oxford University Press, 1979, pp. 70–2.

26 Du Cros, p. 14.

27 H T Tisdall, 'Walhalla as a Collecting Ground', *Victorian Naturalist*, vol. 11, no. 11, February 1895, pp. 147–51.

28 South Gippsland Development League, *The Land of the Lyre Bird: A Story of Early Settlement in the Great Forest of South Gippsland*, Shire of Korumburra, Korumburra, 1920, pp. 31–3, 61.

29 Deborah Bird Rose (ed.) *Country in Flames*, Biodiversity Unit, Department of the Environment, Sport and Territories, and the North Australia Research Unit, Australian National University, Canberra and Darwin, 1995; Peter Latz, *Bushfires and Bushtucker: Aboriginal Plant Use in Central Australia*, Alice Springs, 1995; and J R W Reid, J A Kerle and S R Morton (eds), *Uluru Fauna*, CSIRO, Canberra, 1993.

30 Sue Feary and Gregg Borschmann, 'The first foresters: The archaeology of Aboriginal forest management', in Gregg Borschmann (ed.), *The People's Forest: A Living History of the Australian Bush*, The People's Forest Press, Sydney, 1999, pp. 13–22, at p. 17.

31 Du Cros; Upcher; D Byrne, *The Five Forests*, National Parks and Wildlife Service, Sydney, 1983; Peter Gell and Iain-Malcolm Stuart, *Human Settlement History and Environmental Impact: the Delegate River Catchment, East Gippsland, Victoria*, Monash Publications in Geography no. 36, Department of Geography and Environmental Science, Monash University, Melbourne, 1989.

32 Alfred Howitt, 'The Eucalypts of Gippsland', *Transactions of the Royal Society of Victoria*, vol. 2, no. 1, 1890, pp. 81–120. I have also drawn upon, and benefited from, Bill Gammage's succinct and perceptive discussion of Howitt's paper in his 'Sustainable damage: the environment and the future', in Stephen Dovers (ed.), *Australian Environmental History: Essays and Cases*, Oxford University Press, Melbourne, 1994, 258–67.

33 Howitt, 'The eucalypts of Gippsland', p. 111.

34 Discussion of Howitt's paper can be found in *Proceedings of the Royal Society of Victoria*, vol. 3, 1891, pp. 124–9. I am grateful to Bill Gammage for drawing my attention to this source.

35 Eric Rolls, *A. Million Wild Acres*, Nelson, Melbourne, 1981

36 Eric Rolls, More a New Planet Than a New Continent, CRES public lecture, Canberra, 1985, p. 2.

37 Rolls, *A Million Wild Acres*, p. 399.

38 Rolls has since defended and refined his argument in 'Land of grass: The loss of Australia's grasslands', *Australian Geographical Studies*, vol. 37, 1999, pp. 197–213, and 'The end, or new beginning?', in Stephen Dovers (ed.), *Environmental History and Policy: Still Settling Australia*, Oxford University Press, Melbourne, 2000, pp. 24–46.

39 John Dargavel, *Fashioning Australia's Forests*, Oxford University Press, Melbourne, 1995, pp. 184–5, and Ian Watson, *Fighting over the Forests*, Allen & Unwin, Sydney, 1990.

40 Rolls, *A Million Wild Acres*, p. 402.

41 See Tom Griffiths, 'The writing of *A Million Wild Acres*', in John Dargavel, Di Hart and Brenda Libbis (ed.) *The Perfumed Pineries*, Australian Forest History Society, Canberra, 2001.

42 For example, see Walters, 'Botanical tour of the Dandenong Ranges'; evidence of A F Kelso, in 'Transcript of evidence given before the Royal Commission to enquire into the causes and origins and other matters arising out of bush fires in Victoria during the month of January 1939', vol. 1 of 3 vols, Department of Natural Resources and Environment Library, Melbourne, pp. 110–12; evidence of J A Bostock, in 'Transcript of evidence ... 1939', p. 692; David Brown quoted in Sally Wilde, *Forests Old, Pastures New: A History of Warragul*, Shire of Warragul, Warragul, 1988, p. 10; South Gippsland Development League, *Land of the Lyre Bird*.

3 'IMPROVING'

1 W K Hancock, *Discovering Monaro*, Cambridge University Press, Cambridge, 1972, p. 72.

2 Caroline Philipp, 'Farming at Gulf Station', *Trust News*, November 1989, pp. 12–13.

3 Tony Dingle, *Settling* (vol. 2 of *The Victorians*), Fairfax, Syme & Weldon, Sydney, 1984, p. 25.

4 W B Calder, 'The natural vegetation pattern of the Mornington Peninsula with particular reference to the genus *Eucalyptus*', MSc thesis, University of Melbourne, 1972, p. 307.

5 J Blackburn, *Reports on State Forests, 1885–1911*, especially reports on Mount Fatigue Timber Reserve (7/5/1888) and Forest, Agnes River, South Gippsland (15/4/1891), Department of Natural Resources and Environment, Victoria (drawn to my attention by Paul Barker). For more on the ringbarking debates, see Brett J Stubbs, 'Land improvement or institutionalised destruction? The ringbarking controversy, 1879–1884, and the emergence of a conservation ethic in New South Wales', *Environment and History*, vol. 4, no. 2, June 1998, pp. 145–67.

6 D Hunt, *History of Neerim*, the Author, Melbourne, p. 106.

7 N J Caire, 'Notes on the giant trees of Victoria', *Victorian Naturalist*, 21, 9, (Jan 1905), pp. 122–8.

8 J Blackburn, *Reports on State Forests, 1885–1911*, Department of Natural Resources and Environment.

9 A V Galbraith, 'Memo for the Hon. The Premier' on the first few years of the Forests Commission's Activities, 1919–1922, Forests Commission File no. 24/96, Department of Natural Resources and Environment, Melbourne.

10 Dingle, p. 138.

11 Quoted in W S Noble, *Ordeal by Fire: The Week a State Burned Up*, Jenkin Buxton Printers Pty Ltd, Melbourne, 1977, pp. 10–11.

12 Quoted in G. Bolton, *Spoils and Spoilers: Australians Make their Environment, 1788–1980*, Allen & Unwin, Sydney, 1981, p. 41.

13 South Gippsland Development League, *The Land of the Lyre Bird: A Story of Early Settlement in the Great Forest of South Gippsland*, Shire of Korumburra, Korumburra, 1920, p. 54.

14 *Land of the Lyre Bird*, p. 269.

15 *Land of the Lyre Bird*, p. 59.

16 J B Swaffield, 'Back to Neerim history', quoted in *Warragul and District Historical Society Monthly Bulletin*, no. 48, 17 April 1972.

17 Warwick Frost, 'European farming, Australian pests: Agricultural settlement and environmental disruption in Australia, 1800–1920', *Environment and History*, vol. 4, no. 2, June 1998, pp. 135–6.

18 Quoted in Dingle, *Settling*, p. 67.

19 Ailsa McLeary with Tony Dingle, *Catherine: On Catherine Currie's Diary, 1873–1908*, Melbourne University Press, Melbourne, 1998, pp. 54, 77–8.

20 Dingle, p. 67.

21 Stephen J Pyne, *Burning Bush: A Fire History of Australia*, Henry Holt & Co, New York, 1991, p. 243.

22 Pyne, pp. 240–4.

23 Sally Wilde, *Forests Old, Pastures New: A History of Warragul*, Shire of Warragul, Warragul, 1988, p. 42.

24 J W Audas, *The Australian Bushland*, W A Hamer, North Melbourne, 1950.

25 Alfred W Crosby, *Ecological Imperialism: The Biological Expansion of Europe, 900–1900*, Cambridge University Press, Cambridge, 1986, p. 194.

26 Linden Gillbank, 'The origins of the Acclimatisation Society of Victoria: Practical science in the wake of the gold rush', *Historical Records of Australian Science*, vol. 6, no. 3, December 1986, p. 371; Cecil J le Souef, 'Acclimatisation in Victoria', *Victorian Historical Magazine*, vol. 36, pp. 8–29; *Report from the Select Committee of the Legislative Assembly upon the Zoological and Acclimatisation Society of Victoria Incorporation Bill*, Minutes of Evidence, Victorian government, Melbourne, 19 August 1884, pp. 10–11; Arthur Bentley, *An Introduction to the Deer of Australia with Special Reference to Victoria*, Koetung Trust, Melbourne, 1967, rev. edn 1978, p. 35.

27 Jack Ritchie, 'Introduction of brown trout to Victoria', *The Australian Trout*, Victorian Fly Fishers' Association, Melbourne, 1988, pp. 72ff.

28 Le Souef, pp. 21–2; Bentley, p. 34.

29 Forests Commission, Victoria, *Annual Report of Powelltown Forest District, 1939*, VPRS 10568, Public Record Office, Victoria.

30 J B Swaffield, 'Back to Neerim history'.

31 L E B Stretton, *Report of the Royal Commission to Inquire Into Forest Grazing*, Victorian government printer, Melbourne, 1946, pp. 16–17.

32 Warwick Frost, 'European Farming, Australian Pests: Agricultural Settlement and Environmental Disruption in Australia, 1800–1920', *Environment and History*, vol. 4, no. 2, June 1998, pp. 129–43.

33 *Land of the Lyre Bird*, pp. 114–15.

34 G Spicer to Secretary for Lands, 30 September 1892, VPRS 440, 1319/42.44, Public Record Office, Victoria.

35 Jan Miller and Isabel Buckland, *Warburton Village Settlement*, the authors, Yellingbo, 1987.

36 Thomas Hughes to Secretary of Village Settlements, 24 August 1896, VPRS 5357, 3401/5.10, Public Record Office, Victoria.

37 John Berry to Department of Lands and Survey, 9 July 1897, VPRS 5357, 7992/5.10, Public Record Office, Victoria.

38 Marilyn Lake, *The Limits of Hope: Soldier Settlement in Victoria, 1915–1938*, Oxford University Press, Melbourne, 1987; J M Powell, *An Historical Geography of Modern Australia: The Restive Fringe*, Cambridge University Press, Cambridge, 1988, p. 111 and map on p. 115; Meredith Fletcher, *Avon to the Alps*, Shire of Avon, Stratford, 1988, ch. 6.

39 W D Ingle to the Acting Conservator of Forests, July 1905, VPRS 440, 4036/187, Public Record Office, Victoria.

40 Julie Marshall, 'A history of Keppel's Hut', unpublished report to the Historic Places Branch of the Department of Conservation and Environment, Melbourne, 1991; P B Cabena, 'Grazing the high country: An historical and political geography of high country grazing in Victoria, 1835–1935', MA thesis, University of Melbourne, 1980, pp. 47–8.

41 Evidence of William Francis Lovick, grazier, in 'Transcript of evidence', p. 696.

42 A V Galbraith, 'Forest fires: cause and effect', *The Gum Tree*, 1926, pp. 1–2.

43 State Forests Department of Victoria, *Annual Report 1914*, Melbourne, 1915.

44 Quoted in Noble, p. 12.

45 Evidence given to the 1939 Royal Commission, in 'Transcript of evidence', pp. 676, 697.

46 *Mansfield Courier*, 24 February 1939, quoted in Mark Sainsbery, 'Forestry in the Mansfield District; including the impact of the 1939 Bushfires', unpublished report to the Historic Places Branch of the Department of Conservation and Environment, Melbourne, 1990.

47 Stretton, *Report of the Royal Commission to Inquire Into Forest Grazing*.

48 Grazing lease file of Henry Middleton, VPRS 440, 5054/42, Public Record Office, Victoria.

49 Beetham; Forests Commission of Victoria, Annual Reports of the Neerim Forest District, 1930s, VPRS 10568, Public Record Office, Victoria.

50 Village community allotment 50, Parish of Neerim, VPRS 5357, 7992/5. 10, Public Record Office, Victoria.

51 For example, *Land of the Lyre Bird*, pp. 213–14.

52 *Land of the Lyre Bird*, p. 218.

53 Quoted in Nettie Palmer, *The Dandenongs*, National Press, Melbourne, 1953, pp. 50–1.

4 CROSSING THE BLACKS' SPUR

1 Tim Bonyhady, 'Primeval forests in Australia', in John Dargavel (ed.), *Australia's Ever-Changing Forests III*, Centre for Resource and Environmental Studies, Australian National University, Canberra, 1997, pp. 30–3.

2 Diane Barwick, *Rebellion at Coranderrk*, Laura E Barwick and Richard E Barwick (eds), Aboriginal History Monograph 5, Canberra, 1998, pp. 21–3; Henry Reynolds, *The Law of the Land*, second edition, Penguin, Melbourne, 1992, pp. 125–8; Bain Attwood, 'Batman's Treaties', in Andrew Brown-May (ed.), *Encyclopedia of Melbourne*, forthcoming.

3 Heather Goodall, *Invasion to Embassy: Land in Aboriginal Politics in New South Wales, 1770–1972*, Allen and Unwin, Sydney, 1996, p. 88.

4 Heather Goodall, *Invasion to Embassy*, pp. 76–7.

5 William Thomas, Guardian of Aborigines, to the Commissioner of Lands and Survey, Victoria, 4 March 1959, in Bain Attwood and

Andrew Markus (eds), *The Struggle for Aboriginal Rights: A Documentary History*, Allen & Unwin, Sydney, 1999, pp. 41–3.

6 Quoted by Bain Attwood and Andrew Markus, 'The nineteenth century', in Attwood and Markus, p. 32.

7 Barwick, *Rebellion*, p. 62.

8 William Thomas to R Brough Smyth, 5 October 1860, in Attwood and Markus, p. 43.

9 D J Mulvaney, *Encounters in Place: Outsiders and Aboriginal Australians 1606–1985*, University of Queensland Press, St Lucia, 1988, p. 149; Barwick, Rebellion, p. 65.

10 Barwick, *Rebellion*, p. 65.

11 Quoted in Barwick, *Rebellion*, p. 67.

12 Barwick, *Rebellion*, p. 66; Mulvaney, p. 153.

13 Quoted in Barwick, *Rebellion*, p. 75.

14 Barwick, *Rebellion*, pp. 62, 78. In my account of Coranderrk I have also drawn on 'Report of the Board appointed to inquire into Coranderrk Aboriginal Station', Victoria, Legislative Assembly, *Votes and Proceedings 1882–83*, vol. 2; Barwick, 'Coranderrk and Cumeroogunga: Pioneers and policy', in T Scarlett Epstein and David H Penny (eds), *Opportunity and Response: Case Studies in Economic Development*, C Hurst & Company, London, 1972, pp. 10–68; Bain Attwood, *The Making of the Aborigines*, Allen & Unwin, Sydney, 1989, pp. 92–5; M Goulding, 'Aboriginal occupation of the Melbourne Area, District 2', unpublished report to the Land Conservation Council, Melbourne, 1988, pp. 47–50; Marie Hansen Fels, 'Coranderrk: A Station never a Mission', *Site: The Newsletter of the Victoria Archaeological Survey*, no. 8, Winter 1990, pp. 14–15, and Mulvaney, *Encounters in Place*, pp. 146–57.

15 Mulvaney, *Encounters*; Fels, 'Coranderrk'.

16 Barwick, *Rebellion*, pp. 6, 69.

17 Barwick, *Rebellion*, p. 5.

18 Inga Clendinnen, *True Stories*, Boyer Lectures 1999, ABC Books, Sydney, 1999, p. 80.

19 Quoted in Patricia Marcard, 'William Barak (1824–1903), *Australian Dictionary of Biography*, vol. 3, Melbourne University Press, Melbourne, 1969. See also Barak's evidence in 'Report of the Board into Coranderrk', pp. 9–10.

20 See *The Leader* (Melbourne) Supplement, 15 July 1882, in Attwood and Markus, pp. 46–7, for an 'emphatic denial' by the Kulin of outside influence.

21 Jane Lydon, 'Regarding Coranderrk: Photography at Coranderrk Aboriginal Station, Victoria', PhD thesis, Australian National University, Canberra, 2000.

22 Edward M Curr, evidence, 'Report of the Board into Coranderrk', p. 120.

23 'Report of the Board into Coranderrk', p. 120.

24 Lydon, 'Regarding Coranderrk'; Nicholas Caire,'Notes on the giant trees of Victoria', *The Victorian Naturalist*, vol. 21, no. 9, January 1905, pp. 122–8.

25 Barwick, *Rebellion*, p. 5.

26 'Transcript of evidence given before the Royal Commission to enquire into the causes and origins and other matters arising out of bush fires in Victoria during the month of January 1939', vol. 1 of 3 vols, Department of Natural Resources and Environment Library, Melbourne, pp. 205–7.

27 'Swap of axe for sacred site makes dream reality', *Age*, 6 September 1991.

28 Daryl Tonkin and Carolyn Landon, *Jackson's Track: Memoir of a Dreamtime Place*, Penguin Books, Melbourne, 1999, pp. 6, 30–3, 43–6, 58–60.

29 Alick Jackomos and Derek Fowell (ed.), *Living Aboriginal History of Victoria: Stories in the Oral Tradition*, Museum of Victoria, Aboriginal Cultural Heritage Advisory Committee, Cambridge University Press, Cambridge, 1991, p. 20.

30 Tonkin and Landon, p. 214.

31 Penny Tripcony, 'Towards Aboriginal management of Aboriginal rental housing, Melbourne, 1960–89', in Peter Read (ed.), *Settlement: A History of Australian Indigenous Housing*, Aboriginal Studies Press, Canberra, 2000, p. 147.

32 Tonkin and Landon, p. 254.

33 Tonkin and Landon, pp. 255–70.

5 MINING

1 George Gordon McCrae, 'Black Thursday, 6 February 1851', *La Trobe Library Journal*, vol. 11, no. 44, Spring 1989, pp. 20–1; Rolf Boldrewood, *Old Melbourne Memories*,

William Heinemann, 1969, p. 118, and William Howitt, *Land, Labour and Gold*, Boston, 1855, vol. 2, pp. 190–1, quoted in Pyne, *Burning Bush*, pp. 221–2.

2 Anne V Bailey and Robin A Bailey, 'Matlock: An alpine gold mining town in the 1860s', *Victorian Historical Magazine*, vol. 48, no. 4, November 1977, pp. 247–59.

3 H J Stacpoole, 'The discovery of the Woods Point goldfields', *Victorian Historical Magazine*, vol. 37, pp. 50–72; Brian Lloyd and Howard Combes, *Gold at Gaffneys Creek*, Shoestring Bookshop, Wangaratta, 1981; N Houghton, 'Tramways of Woods Point District, 1863–68', *Light Railways*, Spring 1975, pp. 4–15.

4 'The diary of Captain [Jonathan] Solomon', first published in the *Woods Point Mountaineer* on 9 September 1864, reprinted in H J Stacpoole, *Tracks to the Jordan*, Lowden, Kilmore, 1973, pp. 10–20; Peter Pierce (ed.) *The Oxford Literary Guide to Australia*, Oxford University Press, Melbourne, 1987, entry under 'Jericho'.

5 Anonymous, 'A visit to the Australian Alps' (condensed from an article in *Dicker's Mining Record*, April 1864), in Stacpoole, pp. 45–55.

6 Stacpoole, p.7.

7 Bailey and Bailey.

8 For example, Anthony Trollope, *New South Wales, Queensland, Victoria and Tasmania*, Ward, Lock & Co., London, part II, p. 68.

9 Stacpoole, pp. 70–1.

10 Stacpoole, p. 65.

11 R Brough Smyth, *Progress Report of the Geological Survey of Victoria, 1876*, Victorian Government, Melbourne, 1876.

12 Brian Carroll, *The Upper Yarra: An Illustrated History*, Shire of Upper Yarra, Yarra Junction, 1988, pp. 36–7.

13 Trollope, p. 69.

14 John Adams, *Mountain Gold: A History of the Baw Baw and Walhalla country of the Narracan Shire, Victoria*, Narracan Shire Council, Trafalgar, 1980, p. 69.

15 John Sadleir, *Recollections of a Victorian Police Officer*, G Robertson, Melbourne, 1913, quoted in Pierce, entry under 'Walhalla'; Lou de Prada, 'Reminiscences of a man who was born at Walhalla', typescript,

Centre for Gippsland Studies, Monash University Gippsland, n.d.

16 Trollope, pp. 67–8; Victorian Railways, *Picturesque Victoria and How to Get There: A Handbook for Tourists*, Osboldstone & Co., Melbourne, 1912, p. 225.

17 H T Tisdall, 'Under Eastern Baw Baw: A Botanical Trip in the Gippsland Mountains', *Victorian Naturalist*, vol. 13, no. 7, (Oct 1896), p. 94.

18 De Prada, p. 2.

19 Mark Plummer, 'The Walhalla mining tramways', *Light Railways*, no. 16, pp. 1921; *Walhalla and Thomson River Steam Tramway*, Walhalla, 1972 (pamphlet).

20 De Prada, p. 3.

21 Lloyd and Combes.

22 Peter Evans, 'Wolfram Mine, Marysville', *Light Railway News*, no. 65, August 1988, p. 4; Light Railway Research Society of Australia, *Erica–Walhalla Tour, 27–28 May 1989*, Light Railway Research Society of Australia, Surrey Hills, 1989, pp. 9–10.

23 D W Paterson, 'Woods Point Forest District', *Victorian Foresters' Newsletter*, no. 10, 1958.

24 De Prada, p. 6.

6 TIMBER TRAMWAYS

1 The history of sawmilling in Victoria is described in many of the publications of the Light Railway Research Society of Australia, especially their *Tall Timber and Tramlines: An Introduction to Victoria's Timber Tramway Era*, LRRSA, Surrey Hills, 1974. See also L T Carron, *A History of Forestry in Australia*, ANU Press, Canberra, 1985 (chapter on Victoria); and Gary Vines, 'The historical archaeology of forest based sawmilling in Victoria 1855–1940', BA (Hons) thesis, Department of Archaeology, La Trobe University, Bundoora, 1985.

2 W E Ivey, 'Report upon the Victoria, Dandenong and Bullarook state forests', 1874, in Carver, vol. D, pp. 69–70; Royal Commission on State Forests and Timber Reserves, 1898–1901, *Progress Reports 1–14*, Government Printer, Melbourne.

3 Interview with David Ashton, 2 September 1991.

4　R Wright, *The Bureaucrats' Domain: Space and the Public Interest in Victoria, 1836–84*, Oxford University Press, Melbourne, 1989.

5　Tim Bonyhady, 'Primeval forests in Australia', in John Dargavel (ed.), *Australia's Ever-Changing Forests III*, Centre for Resource and Environmental Studies, Australian National University, Canberra, 1997, p. 30.

6　Ivey; Frederick D'A Vincent, *Notes and Suggestions on Forest Conservancy in Victoria*, Government Printer, Melbourne, 1887; B Ribbentrop, *Report on the State Forests of Victoria*, Victorian Parliamentary Papers, Victorian government, Melbourne, 1896; Royal Commission on State Forests and Timber Reserves, 1898–1901.

7　A V Galbraith, 'Memo for the Hon. The Premier' on the first few years of the Forests Commission's Activities, 1919–1922, Forests Commission File no. 24/96, Department of Natural Resources and Environment, Melbourne.

8　Michael Jones, *Prolific in God's Gifts: A Social History of Knox and the Dandenongs*, Allen & Unwin in association with the City of Knox, Sydney, 1983, p. 91.

9　A W Shillinglaw (Chief, Division of Forest Operations, Forests Commission of Victoria), 'Forests utilization', *Evidence Presented to the State Development Committee on Its Enquiry into the Utilization of Timber Resources in the Watersheds of the State*, Victoria government printer, Melbourne, 1959, pp. 23–9.

10　Vines, p. 41.

11　Galbraith, preface.

12　Vines, p. 121.

13　Evidence of James Francis Ezard, in 'Transcript of evidence', p. 35.

14　The site of Clark & Pearce No. 1 Mill, Rubicon, surveyed by Historic Places (DNRE), 1989.

15　Paul Thornton-Smith, 'Some aspects of the history of Powelltown, 1912–1939', BA (Hons) thesis, Department of History, University of Melbourne, 1976. See also F E Stamford, E G Stuckey and G L Maynard, *Powelltown: A History of Its Timber Mills and Tramways*, Light Railway Research Society of Australia, Surrey Hills, 1984.

16　Alex Larkins, 'River and range: a story of a Victorian river valley', unpublished manuscript (Box 3075/1, MS 12377), La Trobe Collection, State Library of Victoria. Larkins' memoir has fictional names but is based on his experiences in Warburton.

17　Larkins, 'River and range'.

18　'Horty's story', *Tirra Lirra*, Summer 1990–91, pp. 38–41, 59.

19　Ian F McLaren, 'Clarence Michael James Dennis (1876–1938)', *Australian Dictionary of Biography*, vol. 8, Melbourne University Press, Melbourne, 1981; Margaret Herron, *Down the Years*, Hallcraft Publishing Co., Melbourne, 1953.

20　Ian D Rae, 'Wood distillation in Australia: adventures in Arcadian chemistry', *Historical Records of Australian Science*, vol. 6, no. 4, 1987, pp. 469–84; Cuming, Smith & Co., *A Chemical Industry in a Victorian Forest*, Cuming, Smith & Co., Melbourne, n.d.; A P Winzenried, *Britannia Creek: An Essay in Wood Distillation*, APW Productions, Belgrave, 1986.

21　State Electricity Commission, *Power and Heat: Victoria's National Scheme of Electricity and Briquette Production*, SEC, Melbourne, 1928.

22　Gregg Borschmann, *The People's Forest: A Living History of the Australian Bush*, The People's Forest Foundation Ltd, 1999, pp. 250–3.

23　H Ingram, *Early Forest Utilisation*, Forest Recollections series, Institute of Foresters of Australia, Victorian Division, Mitcham, 1979.

24　Ingram; see also John Dargavel, *Sawing, Selling and Sons: Histories of Australian Timber Firms*, Centre for Resource and Environmental Studies, Canberra, 1988; Mike McCarthy, *Trestle Bridges and Tramways: The Timber Industry of Erica District 1910–1950*, Light Railway Research Society of Australia, Surrey Hills, 1983.

25　Forests Commission, Victoria, Annual Reports of the Upper Yarra Forest District, 1931–39, VPRS 10568, Public Record Office, Victoria.

26　M Dixson, 'The timber strike of 1929', *Historical Studies*, vol. 10, no. 40, 1963, pp. 479–92.

27　Larkins.

28　Forests Commission, Victoria, Annual Report of the Upper Yarra Forest District, 1931.

29　Thornton-Smith, pp. 29–30.

30 Stamford, Stuckey and Maynard, p. 111; Annual Reports of the Neerim Forest District, 1930s.

31 Larkins.

7 WATER

1 In 'Transcript of evidence ... 1939', pp. 91–128.

2 'Transcript of evidence ... 1939', pp. 99, 101–2.

3 Patrick O'Shaughnessy, *Melbourne and Metropolitan Board of Works Catchment Management Policies: A History and Analysis of their Development*, CRES working paper 1986/28, ANU Canberra, 1986, p.1.

4 Interviews with Richard Marchant, Museum Victoria, 8 June 1999 and 6 April 2000; P S Lake and R Marchant, 'Australian upland streams: ecological degradation and possible restoration', *Proceedings of the Ecological Society of Australia*, vol. 16 (1990), pp. 79–91.

5 O'Shaughnessy, p.1.

6 R C Seeger, 'The history of Melbourne's water supply – Part 1', *Victorian Historical Magazine*, vol. 19, pp. 107–19, 133–8; J M Powell, *Environmental Management in Australia, 1788–1914*, Oxford University Press, Melbourne, 1976, pp. 44–5.

7 Tony Dingle, 'The Cascades', in Tom Griffiths, *Secrets of the Forest*, Allen & Unwin, Sydney, 1992, pp. 178–80.

8 Dingle, 'The Cascades'.

9 Tony Dingle and Carolyn Rasmussen, *Vital Connections: Melbourne and its Board of Works 1891–1991*, McPhee Gribble, Melbourne, 1991, p. 32.

10 J C Jessop, *A History of the MMBW*, Melbourne and Metropolitan Board of Works, Melbourne, 1942, p. 23; Dingle and Rasmussen, p. 400, appendix 3: 'Water consumption graph'.

11 Dingle and Rasmussen, pp. 368–70.

12 Susan Priestley, *Making Their Mark* (vol. 3 of *The Victorians*), Fairfax, Syme & Weldon, Sydney, 1984, p. 326.

13 Dingle and Rasmussen, pp. 220–1.

14 O'Shaughnessy; John Knowles, 'Board of Works closed catchment policy, 1890–1960',

assignment for the Master of Environmental Science program (June 1989), Monash University, 1989.

15 O'Shaughnessy, p. 6.

16 Jessop, p. 26.

17 O'Shaughnessy.

18 Dingle and Rasmussen, pp. 114–15, 267.

19 O'Shaughnessy; Knowles.

20 H Ingram, *Early Forest Utilisation*, Forest Recollections series, Institute of Foresters of Australia, Victorian Division, Mitcham, 1979, p. 6.

21 Dingle and Rasmussen, pp. 190–2, 258–61.

22 Quoted in L T Carron, *A History of Forestry in Australia*, ANU Press, Canberra, 1985, pp. 181–2.

23 O'Shaughnessy.

24 Dingle and Rasmussen, p. 261.

25 J D Brookes, 'The relation of vegetation cover to water yield in Victorian mountain watersheds', MSc thesis, University of Melbourne, 1949; interview with David Ashton, 2 September 1991; Ashton, 'Fire in tall open-forests', p. 346.

26 This is A E Kelso's description, in 'Transcript of evidence', p. 108.

27 Interview with David Ashton, 2 September 1991.

28 O'Shaughnessy, p. 21.

29 Dingle and Rasmussen, p. 370.

30 Dingle and Rasmussen, pp. 368–75. On the history of water management in Victoria, see J M Powell, *Watering the Garden State: Water, Land and Community in Victoria, 1834–1988*, Allen & Unwin, Sydney, 1989.

8 THE THEATRE OF NATURE

1 Baldwin Spencer, 'Report of a visit to the Yarra Falls', *Victorian Naturalist*, 7, 11 (March 1891), pp. 157–79.

2 Geo. Lyell, 'To Yarra Falls in the eighties', *Victorian Naturalist*, 45, 8 (Dec 1928), pp. 203–6.

3 Barnard, *ibid*, pp. 250–1.

4 Ross Field, 'Melbourne's Butterflies – MacLeay's Swallowtail, *Graphium macleayanus*', Museum Victoria, 2000.

5 W H A Roger's notes in F G A Barnard, 'Excursion to Launching Place', *Victorian Naturalist*, 25, 1 (May 1908), p. 9.

6 A D Hardy and Mrs Hardy, 'A Tramp from Healesville to Buxton', *Victorian Naturalist*, 22, 10 (Feb 1906), pp. 163–74.

7 David Elliston Allen, *The Naturalist in Britain: A Social History*, London, Allen Lane, 1976, pp. 152–3.

8 F G A Barnard, 'In the Valley of the Upper Yarra', *Victorian Naturalist*, 23, 12 (April 1907), pp. 245–52.

9 Arthur Dendy, 'Zoological notes on a trip to Walhalla', *Victorian Naturalist*, 6, 7 (Nov 1889), pp. 128–36.

10 H T Tisdall, 'Walhalla as a collecting ground', *Victorian Naturalist*, 11, 11 (Feb 1895), pp. 147–51.

11 H T Tisdall, 'Fungi of country east of Mt Baw Baw', *Victorian Naturalist*, 1 (1884–85), pp. 169–72.

12 F G A Barnard, 'The "Camp-Out" at Maroondah Weir', *Victorian Naturalist*, 17, 8 (Dec 1900), pp. 131–8.

13 A D Hardy, 'The Plenty Ranges in Early Spring', *Victorian Naturalist*, 24, 8 (Dec 1907), pp. 128–34, at p. 132.

14 A D Hardy, 'The Plenty Ranges in Early Spring', p. 134.

15 'The "Camp Out" at Olinda Creek', *Victorian Naturalist*, vol. 1 (1884–85), pp. 110–12.

16 Land Conservation Council, *Melbourne Area District 2 Review: Final Recommendations*, LCC, Melbourne 1994, p. 96.

17 A J Campbell, *Nests and Eggs of Australian Birds*, quoted in Charles Barrett, 'Haunts of the Helmeted Honeyeater', *Victorian Naturalist*, 50, 7 (Nov 1933), p. 162.

18 J A Kershaw, 'Excursion to Lower Ferntree Gully', *Victorian Naturalist*, 15, 10 (Feb 1899), pp. 124–6.

19 A J Campbell, 'Some Australian birds – lyrebirds', *The Australasian*, 5 December 1896.

20 Tom Griffiths, *Hunters and Collectors: The antiquarian imagination in Australia*, Cambridge University Press, Cambridge, 1996, pp. 12–21.

21 Charles Barrett, *Koonwarra: A Naturalist's Adventures in Australia*, London, Oxford University Press, 1939, p. 35.

22 'The Woodlanders' [Charles Barrett], 'Our Bush Hut on Olinda', Parts 1–5, *The New Idea: A Woman's Home Journal for Australasia*, October, November, December, 1905, January, February 1906. The quote is from the issue of 6 October 1905, p. 354. See also C L Barrett and E B Nicholls, 'Bird notes from Olinda Vale', *Victorian Naturalist*, 21, 11 (March 1905), pp. 162–6; C L Barrett, 'Bird life on Olinda Creek', *Victorian Naturalist*, 23, 4 (Aug 1906), pp. 84–9, and E B Nicholls, 'Excursion to Olinda Vale', *Victorian Naturalist*, 23, 10 (Feb 1907, pp. 172–3.

23 Barrett, *Koonwarra*, p. 37.

24 Barrett, 'Haunts of the Helmeted Honeyeater', pp. 163–5.

25 Hudson is quoted in P Morton, *The Vital Science: Biology and the Literary Imagination, 1860–1900*, George Allen & Unwin, London, 1984, pp. 70–75; 'The Woodlanders', *New Idea*, 6 January 1906, pp. 666–70.

26 Charles Barrett, *From Range to Sea: A Bird Lover's Ways*, Melbourne, T C Lothian, 1907, pp. 21, 25, 34.

27 The first restrictions on egg collecting during RAOU camps were enforced in 1923, and it was finally prohibited in 1933: D L Serventy, 'A historical background of ornithology with special reference to Australia', *The Emu*, vol. 72, part 2, April 1972, p. 48.

28 Michael Sharland, 'Memories of Tom Tregellas', *The Australian Bird Watcher*, 9, 4, December 1981, pp. 103–9.

29 Tom Tregellas, 'The truth about the lyrebird', *Emu* 30 (1930), pp. 243–50.

30 Tregellas, 'The truth about the lyrebird', p. 249.

31 L H Smith, 'The superb lyrebird *Menura novaehollandiae* – A comment on single-species management', *Australian Bird Watcher*, 16, 4, December 1995, pp. 169–71.

32 A H Chisholm, 'Memorable days near Melbourne', *Victorian Naturalist*, 50, 6 (Oct 1933), pp. 143–6.

33 L H Smith, 'A critical analysis of the factors responsible for the decline of the superb lyrebird *Menura noveahollandiae* in Sherbrooke Forest, Victoria', *Australian Bird Watcher*, 15, 6, June 1994, pp. 238–49.

34 Janie Kirkpatrick, *A Continent Transformed: Human Impact on the Natural Vegetation of*

Australia, Oxford University Press, Melbourne, 1994, p. 33.

35 L H Smith, *The Life of the Lyrebird*, William Heinemann Australia, Melbourne, 1988, p. xiv.

9 TOURISM

1 Betterment and Publicity Board, Victorian Railways, *Melbourne's Nearer Ranges* (pamphlet), November 1929, p. 1 (kindly lent by Sue Stanley, Launching Place).

2 J P O'Meara [?], 'Diary of a tour to the country, 1889', unpublished manuscript, La Trobe Collection, State Library of Victoria.

3 This was the opinion of Melbourne journalist, the 'Vagabond', quoted in Sally Symonds, *Healesville: History in the Hills*, Pioneer Design Studio, Lilydale, 1982, p. 62.

4 *Illustrated Australian News*, 1872, quoted in Elizabeth Johns, Andrew Sayers, Elizabeth Mankin Kornhauser with Amy Ellis, *New Worlds from Old: 19th Century Australian and American Landscapes*, National Gallery of Australia, Canberra and Wadsworth Atheneum, Hartford, Connecticut, National Gallery of Australia, Canberra, 1998, p. 181.

5 Geoffrey Serle, *John Monash: A Biography*, Melbourne University Press, Melbourne, 1982, p. 70.

6 A W Howitt to Anna Mary Howitt, 15 January 1858, quoted in Tim Bonyhady, *The Colonial Earth*, Melbourne University Press, Melbourne, 2000, p. 107.

7 Tim Bonyhady, *Images in Opposition: Australian Landscape Painting, 1801–1890*, Oxford University Press, Melbourne, 1985, pp. 64–7, and *The Colonial Earth*, chapter 4.

8 N J Caire and J W Lindt, *Companion Guide to Healesville, Blacks' Spur, Narbethong and Marysville*, Atlas Press, Melbourne, 1904, p. 34; A and D Pitkethly, *N J Caire: Landscape Photographer*, published by the authors, Rosanna, 1988; S Jones, *J W Lindt: Master Photographer*, Currey O'Neil Ross on behalf of the Library Council of Victoria, South Yarra, 1985.

9 Victorian Railways.

10 W F Waters, 'The Baw Baws', *The Melbourne Walker*, vol. 37, Melbourne Walking Club, Melbourne, 1966.

11 R H Croll, *I Recall: Collections and Recollections*, Robertson and Mullens Ltd., Melbourne, 1939, p. 81.

12 Harry Stephenson, 'The early days', *Wild*, issue 16, Autumn 1985, p.48.

13 *Sydney Mail*, 29 July 1931, quoted in Melissa Harper, 'The Battle for the Bush: Bushwalking vs Hiking Between the Wars', *Journal of Australian Studies*, no. 45 (June 1995), p. 41.

14 Quoted in Harper, p. 44.

15 Some of Croll's suggested walks were reproduced in Betterment and Publicity Board, Victorian Railways, *Wonderful Walks in Victoria*, Australia (pamphlet), June 1927 (kindly lent by Sue Stanley, Launching Place).

16 *Tourists Map of Narbethong and Marysville Districts*, Victoria, Department of Lands and Survey, Melbourne, 1913.

17 Paul Thornton-Smith, 'Some aspects of the history of Powelltown, 1912–1939', BA (Hons) thesis, Department of History, University of Melbourne, 1976, pp. 13–14.

18 Mike McCarthy, *Bellbrakes, Bullocks and Bushmen: A Sawmilling and Tramway History of Gembrook, 1885–1985*, Light Railway Research Society of Australia, Melbourne, 1987, p. 27.

19 F G A Barnard, 'Excursion to Warburton', *Victorian Naturalist*, 22, 8 (Dec 1905), pp. 128–32.

20 W F Waters, 'The Gippsland Mountain Goldfields', *The Melbourne Walker*, vol. 35, Melbourne Walking Club, Melbourne, 1964.

21 Stephenson, p. 46.

22 Stephenson, p. 46.

23 For example, see Croll, p. 84.

24 Amy Eastwood, Isabel Eastwood and Hazel Merlo, *Uphill after Lunch: Melbourne Women's Walking Club*, MWWC, Melbourne, c. 1986, pp. 6, 8, 11, 12.

25 Symonds.

26 Anne Hartshom and Laurie Pelech, 'Warrandyte – the forgotten hostel', *The Hosteller*, Summer, 1984–85.

27 Alan D Mickle, 'Healesville', unpublished typescript [no date], MS 8697, Box 2406/6, La Trobe Collection, State Library of Victoria.

28 Symonds, ch. 7.

29 David Fleay, 'Further notes on the Badger Creek platypuses', *Victorian Naturalist*, 67, 4

(1950), p. 81; 'Fauna for export', *Herald*, 22 June 1921; Jane Lydon, 'Regarding Coranderrk', PhD thesis, Australian National University, Canberra, 2000, pp. 199–200.

30 Susan Priestley surveys changing recreational fashions in Victoria in *Making Their Mark* (vol. 3 of *The Victorians*), Fairfax, Syme & Weldon, Sydney, 1984.

31 W C Kernot, 'Melbourne's great unknown mountain', *Victorian Geographical Journal*, vol. 25, 1907, pp. 15–19.

32 Alex Larkins, 'River and range: a story of a Victorian river valley', unpublished manuscript (Box 3075/1, MS 12377), La Trobe Collection, State Library of Victoria, pp. 187, 193.

33 Elaina Fraser, 'We skiied down Mt Erica', *The Gippsland Writer*, vol. 2, no. 1, Winter 1987, pp. 33–5.

34 D Johnson, *The Alps at the Crossroads*, Victorian National Parks Association, Melbourne, 1974, pp. 152–3; Land Conservation Council, p. 226.

35 Sandra Bardwell, 'National parks in Victoria, 1866–1956: For all the people for all time', PhD thesis, Monash University, 1974.

36 Sandra Bardwell, *Fern Tree Gully National Park: A Centenary History, 1882–1982*, National Parks Service, Victoria, Melbourne, 1982.

37 Ian D Lunt, 'Two hundred years of land use and vegetation change in a remnant coastal woodland in Southern Australia', *Australian Journal of Botany*, 46 (1989), pp. 629–47.

10 BLACK FRIDAY

1 For an outstanding fire history of Australia see Stephen J Pyne, *Burning Bush: A Fire History of Australia*, Henry Holt & Co., New York, 1991.

2 Evidence of James Francis Ezard, in 'Transcript of evidence … 1939', p. 32.

3 In this and subsequent paragraphs, I have drawn on W S Noble's account of the fire in *Ordeal by Fire: The Week a State Burned Up*, Jenkin Buxton Printers Pty Ltd, Melbourne, 1977. Noble was a journalist with the Melbourne *Herald* at the time. Other sources on which I have drawn for this account of the fire are L E B Stretton, *Report of the Royal Commission into the Causes of and Measures Taken to Prevent the Bush Fires of Jan. 1939*, Government Printer, Melbourne, 1939, and the associated 'Transcript of evidence' (3 bound volumes of typescript held in the DNRE Library, Melbourne), as well as reports in the *Argus, Age,* and *Herald* of January 1939. See also Jan McDonald, 'Illuminated by a pall of smoke: Victoria and the January 1939 Bushfires', BA (Hons) thesis, Australian National University, 1989.

4 Alex Demby, quoted in the Toolangi Forest Discovery Centre display, Toolangi, 2000.

5 *Sun*, 10 January 1939, p. 3.

6 A moving account of the fire in the Rubicon forest is given by Ernie Le Brun in a video produced by Peter Evans for the Alexandra Timber Tramway and Museum entitled *Rails to Rubicon*, Melbourne, 1989. See also Evans' book, *Rails to Rubicon*.

7 Newspaper report quoted by Laurie Duggan, *The Ash Range*, Pan Books, Sydney, 1987.

8 *Herald*, 14 January 1939, p. 1.

9 Evidence of Gerald Alipius Carey, in 'Transcript of evidence', pp. 753–6.

10 'The Angel of Noojee', *Herald*, 11 November 1967, p. 25.

11 Paul Thornton-Smith, 'Some aspects of the history of Powelltown, 1912–1939', BA (Hons) thesis, Department of History, University of Melbourne, 1976, p. 31.

12 Peter Evans, 'Refuge from fire: Sawmill dugouts in Victoria', in John Dargavel (ed.) *Australia's Ever-Changing Forests III*, Centre for Resource and Environmental Studies, Australian National University, Canberra, 1997, p. 222.

13 Quoted in Peter Evans, 'Refuge from fire: sawmill dugouts in Victoria', p. 216.

14 Midgley Ogden, 'My life in the forests of Victoria and the timber industry (1914–1976)', unpublished typescript, Melbourne, p. 133, Department of Natural Resources and Environment Library. Part of this account has been published in Dargavel, *Sawing, Selling and Sons*, Centre for Resource and Environmental Studies, 1988, pp. 93–9.

15 'Gifts of tobacco wanted', *Age*, 25 January 1939, p. 5.

16 Stretton, p. 5.

17 A copy of this film (produced in association with 20th Century Fox) is held on video in the Historic Places Branch, Department of Natural Resources and Environment, Victoria.

18 Stretton, p. 10.

19 Evidence of Alex Hubert Outhwaite, in 'Transcript of evidence', p. 79.

20 William Howitt, *Land, Labour and Gold*, vol. 2, Longman, Brown, Green and Longmans, London, 1855, p. 156, quoted in Pyne, *Burning Bush*, p. 196.

21 Evidence of James Francis Ezard, in 'Transcript of evidence', p. 32.

22 Evidence of James Francis Ezard, pp. 26, 30.

23 Evidence of Alex Hubert Outhwaite, p. 79.

24 Quoted in Noble, p. 10.

25 Pyne, *Burning Bush*.

26 George Perrin, *Report of the Conservator of Forests for the Year Ending 30 June 1890*, Government Printer, Melbourne, 1890, pp. 15–16.

27 C J Dennis's account of the Black Sunday 1926 bushfires is in Herron, p. 61.

28 Forests Commission, Victoria, *Annual Report of the Cann Valley Forest District, 1949*, VPRS 10568, Public Record Office, Victoria.

29 D M Thompson, 'Forest fire prevention and control in the Cann Valley Forest District', Diploma of Forestry thesis, Melbourne, 1952, especially the introduction.

30 'Transcript of evidence', p. 353; *Alexandra and Yea Standard*, 17 February 1939.

31 Evidence of Peter O'Mara, 'Transcript of evidence', p. 1133.

32 'Transcript of evidence', p. 1102.

33 'Transcript of evidence', pp. 353, 425; Memo to R S Code, Inspector of Forests, Ballarat, from G Cockburn, Secretary, Forests Commission, 8 March 1939, attaching an outline by Counsel (Department of Natural Resources and Environment, Victoria, drawn to my attention by Paul Barker).

34 *Sun*, 21 January 1939, editorial.

35 L E B Stretton, 'Remember Black Friday!', *The Riverlander*, no. 66, January 1952, p. 3.

36 'Transcript of evidence', p. 1361.

37 'Transcript of evidence', pp. 1778, 153.

38 'Transcript of evidence', *passim*. As an example, see the evidence of A E Kelso and of C E Lane-Poole.

39 L E B Stretton, 'Judge Stretton's reminiscences', *La Trobe Library Journal*, vol. 5, no. 17, April 1976.

40 The anthology is *Land of Wonder*, ed. A H Chisholm, Angus & Robertson, Sydney, 1964.

41 Stretton, 1946, p. 6.

42 Stretton, 1939, p. 5.

43 Stephen Mark Legg, 'The location of the log-sawmilling industry in Victoria, 1939–77', MA prelim. thesis, Department of Geography, Monash University, 1977, ch. 3.

44 H E Wilkinson, 'The rediscovery of Leadbeater's Possum', *Victorian Naturalist*, vol. 78–9, August 1961, pp. 97–102.

45 David Lindenmayer, *Wildlife and Woodchips: Leadbeater's Possum, A Test Case for Sustainable Forestry*, UNSW Press, Sydney, 1996, pp. 14–15.

46 D R Milledge, C L Palmer and J L Nelson, '"Barometers of Change": The distribution of large owls and gliders in Mountain Ash forests of the Victorian Central Highlands and their potential as management indicators', in Daniel Lunney (ed.), *Conservation of Australia's Forest Fauna*, Royal Zoological Society of NSW, Mosman, 1991, pp. 53 65; Interview with David Ashton, 2 September 1991; Land Conservation Council, pp. 319–21.

47 Forests Commission, Victoria, *Annual Report of the Upper Yarra Forest District, 1940*, VPRS 10568, Public Record Office, Victoria.

48 Forests Commission, Victoria, Fire Protection Officer, *Annual Report for 1939/40*, VPRS 10568/2/40/1550, Public Record Office, Victoria.

49 Evidence of A E Kelso, in 'Transcript of evidence', p. 102.

50 Evidence of Maurice Medlone Dyer (then vice president of the Hardwood Millers' Association), in 'Transcript of evidence', pp. 66–7.

51 Evidence of A E Kelso, pp. 116–17.

52 *Herald* (Melbourne), 16 January 1939, p. 8.

53 H G Wells, *Travels of a Republican Radical In Search of Hot Water*, Penguin Books, Middlesex, England, 1939, chapter IV ('Bush Fires'). I am grateful to Barry Smith for finding this rare book. See his 'H.G. Wells in Australia', *Australian Book Review*, June 2001.

54 Forests Commission, Victoria, *Annual Report of the Upper Yarra Forest District, 1941*, VPRS 10568, Public Record Office, Victoria.

55 Forests Commission, Victoria, *Annual Report of the Erica Forest District, 1941*, VPRS 10568, Public Record Office, Victoria.

56 Legg, p. 34.

57 Forests Commission, Victoria, *Annual Report of the Broadford Forest District, 1940*, VPRS 10568, Public Record Office, Victoria.

58 Forests Commission, Victoria, *Annual Report of the Upper Yarra Forest District, 1952*, VPRS 10568, Public Record Office, Victoria.

59 L T Carron, *A History of Forestry in Australia*, ANU Press, Canberra, 1985, p. 195.

60 Forests Commission, Victoria, *Annual Report of the Erica Forest District, 1950*, VPRS 10568, Public Record Office, Victoria.

61 Forests Commission, Victoria, *Annual Report of the Forests Commission, Victoria 1953–54*, Victorian government printer, Melbourne, 1954; Legg, ch. 4.

62 Ogden, p. 27.

63 Forests Commission, Victoria, *Annual Report of the Forests Commission, Victoria, 1940–41*, Victorian government printer, Melbourne, 1941.

11 THE LONG EXPERIMENT

1 Tim Bonyhady, 'Primeval forests in Australia', in John Dargavel (ed.), *Australia's Ever-Changing Forests III*, Centre for Resource and Environmental Studies, Australian National University, Canberra, 1997, p. 30.

2 Quoted in L T Carron, *A History of Forestry in Australia*, ANU Press, Canberra, 1985, p. 192.

3 A V Galbraith, *Mountain Ash: A General Treatise on Its Silviculture, Management and Utilisation*, Forests Commission of Victoria, Melbourne, 1937, p. 37.

4 A H Beetham, 'Aspects of forest practice in the regenerated areas of the upper Latrobe Valley', Diploma of Forestry thesis, 1950, Department of Natural Resources and Environment Library, p. 2.

5 W E Ivey, 'Report upon the Victoria, Dandenong and Bullarook State Forests', 1874, in M. Carver, 'Forestry in Victoria 1838–1919', vol. D of 5 vols, unpublished typescript, Department of Natural Resources and Environment Library, Melbourne, [no date], p. 75.

6 John Dargavel and Heather McRae, 'Age and order in Victoria's forests', in John Dargavel (ed.), *Australia's Ever-Changing Forests III*, Centre for Resource and Environmental Studies, Australian National University, Canberra, 1997, pp. 67–8.

7 J D Gillespie, 'A survey of denuded forest lands of the Upper Yarra Forest District and a discussion of methods which may be suitable for reforestation of these areas', Diploma of Forestry thesis, Melbourne, 1962, section 1, ch. 6.

8 Daniel Lunney and Chris Moon, 'An ecological view of the history of logging and fire in Mumbulla State forest on the south coast of New South Wales', in *Australia's Ever Changing Forests*, eds K J Frawley and N Semple, Australian Defence Force Academy, Canberra, 1988, p. 35.

9 Forests Commission, Victoria, *Annual Report of the Upper Yarra Forest District, 1937*, VPRS 10568, Public Record Office, Victoria.

10 Beetham, p. 1.

11 Gillespie, 'A survey of denuded forest lands'.

12 The following description of Ashton's work is based on an interview I conducted with him on 2 September 1991, a talk he gave to the Australian Systematic Botany Society on 7 November 1990 entitled 'Wallaby Creek 40 years on', and his article 'A personal look at mountain ash over nearly a life-time', *Tyalla* (Creswick Forestry Students' Magazine), 1987, pp. 3–5. See also D H Ashton, 'The Big Ash forest, Wallaby Creek, Victoria – changes during one lifetime', *Australian Journal of Botany*, vol. 48 (2000), pp. 1–26.

13 J S Turner, 'Control burning and conservation', in *Third Fire Ecology Symposium*, Graduate School of Environmental Science, Monash University, 1974, p. 8.

14 Ashton, 'Fire in tall open-forests'.

15 Ashton, 'The Big Ash forest', pp. 1, 22.

16 See, for example, the forester quoted in Ian Watson, *Fighting over the Forests*, Allen & Unwin, Sydney, 1990, p. 76, the sentiments of the contributors to *Early Forest Utilisation*, and the preface to Carron.

17 Libby Robin, *Building a Forest Conscience: An Historical Portrait of the Natural Resources Conservation League of Victoria (NRCL)*, NRCL, Springvale, 1991, ch. 5.

18 Ingram's and Youl's talks were published in *Early Forest Utilisation*, Forest Recollections series, Institute of Foresters of Australia, Victorian Division, Mitcham, 1979.

19 A V Galbraith, 'Memo for the Hon. The Premier' on the first few years of the Forests Commission's Activities, 1919–1922, Forests Commission File no. 24/96, Department of Natural Resources and Environment, Melbourne.

20 Robin, chapter 1.

21 Robin, p.7.

22 Forests Commission of Victoria film (undated) entitled *Timber* (copy held by the author, kindly supplied by Peter Evans).

23 Pyne, conference paper, p.7.

24 John Dargavel, *Fashioning Australia's Forests*, Oxford University Press, Oxford, 1995, pp. 67–8. See also J M Powell, *An Historical Geography of Modern Australia: The Restive Fringe*, Cambridge University Press, Cambridge, 1988, pp. 33–41.

25 V and R Routley, *The Fight for the Forests: The Takeover of Australian Forests for Pines, Woodchips, and Intensive Forestry*, Research School of Social Sciences, Australian National University, Canberra, 1974; Watson, *Fighting over the Forests*.

26 P Gooday, P Whish-Wilson, and L Weston, 'Regional Forest Agreements: Central Highlands of Victoria', in Australian Bureau of Agricultural Economics, *Australian Forest Products Statistics September Quarter 1997*, Canberra, 1998 (drawn to my attention by Judy Clark). See also Judy Clark, *Australia's Plantations: Industry Employment Environment*, A Report to the State Conservation Councils, Environment Victoria, Melbourne, 1995.

27 Turner; A M Gill and A B Costin, 'Fire and its place in park and wilderness management', in *Third Fire Ecology Symposium*, pp. 12–17; A M Gill, 'Post-settlement fire history in Victorian landscapes', in *Fire and the Australian Biota*, eds A M Gill, R H Groves and I R Noble, Australian Academy of Science, Canberra, 1981, pp. 77–98.

28 See, for example, Peter Gell and Iain-Malcolm Stuart, *Human Settlement History and Environmental Impact: The Delegate River Catchment, East Gippsland, Victoria*, Monash Publications in Geography no. 36, Department

of Geography and Environmental Science, Monash University, Melbourne, 1989.

29 Stephen J Pyne, *Burning Bush*, pp. 410–19.

30 Peter M Attiwill, 'Ecological disturbance and the conservative management of eucalypt forests in Australia', *Forest Ecology and Management*, vol. 63, 1994, pp. 301–46, especially pp. 326, 328; Kay Ansell, 'Risen from the ashes', *Age*, 16 February 1993, p. 11; Graeme O'Neill, 'From ashes to ash', *Time*, 6 September 1993, pp. 58–9.

31 Libby Robin, *Defending the Little Desert: The Rise of Ecological Consciousness in Australia*, Melbourne University Press, Melbourne, 1998, pp. 60–2; also Robin, 'The rise of ecological consciousness in Victoria: the Little Desert dispute, its context and consequences', PhD thesis, University of Melbourne, 1993.

32 Attiwill, 'Ecological disturbance'; O'Neill, 'From ashes to ash'.

33 Robin, *Defending the Little Desert*, pp. 60–2, 72–5, 79, 148–9.

34 Graeme O'Neill and Peter Attiwill, 'Getting ecological paradigms into the political debate: Or will the messenger be shot?', in S T A Pickett, R S Ostfeld, M Shachak and G E Likens (eds), *The Ecological Basis of Conservation: Heterogeneity, Ecosystems, and Biodiversity*, International Thomson Publishing, New York, c. 1997, p. 352.

35 O'Neill, 'From ashes to ash', p. 59.

36 O'Neill and Attiwill, p. 355; O'Neill, 'From ashes to ash'.

37 Interview with David Lindenmayer, Canberra, 18 May 2001; David B Lindenmayer, *Wildlife and Woodchips: Leadbeater's Possum as a test-case of sustainable forestry*, University of NSW Press, Sydney, 1996; Lindenmayer and J F Franklin, 'Re-inventing forestry as a discipline – a forest ecology perspective', *Australian Forestry*, vol. 60, 1997, pp. 53–5; D B Lindenmayer, T W Norton, and M T Tanton, 'Differences between the effects of wildfire and clearfelling in mountain ash forests of Victoria and its implications for fauna dependent on tree hollows', *Australian Forestry*, vol. 53, 1991, pp. 61–8; D B Lindenmayer, B G Mackey, I C Mullen, *et al*, 'Factors affecting stand structure in forests – are there climatic and topographic determinants?', *Forest Ecology and Management*, no. 123, 1999, pp. 55–63;

D B Lindenmayer and J F Franklin, 'Managing stand structure as part of ecologically sustainable forest management in Australian mountain ash forests', *Conservation Biology*, vol. 11, no. 5, October 1997, pp. 1053–68; D B Lindenmayer, R B Cunningham *et al*, 'Structural features of old-growth Australian montane ash forests', *Forest Ecology and Management*, no. 134, 2000, pp. 189–204; D B Lindenmayer, 'Using environmental history and ecological evidence to appraise management regimes in forests', in Stephen Dovers (ed.), *Environmental History and Policy: Still Settling Australia*, Oxford University Press, Melbourne, 2000, pp. 74–96; D B Lindenmayer and E Beaton, *Life in the Tall Eucalypt Forests*, Reed New Holland, Sydney, 2000; M A McCarthy and D B Lindenmayer, 'Multi-aged mountain ash forest, wildlife conservation and timber harvesting', *Forest Ecology and Management*, no. 104, 1998, pp. 43–56; and M A McCarthy, A M Gill and D B Lindenmayer, 'Fire regimes in mountain ash forest: evidence from forest age structure, extinction models and wildlife habitat', *Forest Ecology and Management*, no. 124, 1999, pp. 193–203.

38 D B Lindenmayer, 'Using environmental history and ecological evidence to appraise management regimes in forests', in Stephen Dovers (ed.), *Environmental History and Policy*, p. 75.

39 D B Lindenmayer and E Beaton, *Life in the Tall Eucalypt Forests*, pp, 85, 88.

40 D B Lindenmayer and J F Franklin, 'Managing Stand Structure', p. 1060.

41 D B Lindenmayer and J F Franklin, 'Managing Stand Structure', p. 1053.

42 David Lindenmayer, 'Still searching to save the wood *and* the trees', *Canberra Times*, 24 May 2001, p. 10: 'unfortunately … forest debate has not matured in accordance with current scientific knowledge'.

12 HERITAGE

1 W H C Holmes, 'Scrub cutting', in South Gippsland Development League, *The Land of the Lyre Bird*, Shire of Korumburra, Korumburra, 1920, pp. 54–66.

2 Nettie Palmer, *The Dandenongs*, National Press, Melbourne, 1953, pp. 11–15.

3 *Corhanwarrabul* (Mount Dandenong Historical Society Newsletter), vol. 1, no. 1, 16 October 1974, p. 3.

4 Frederick D'A Vincent, *Notes and Suggestions on Forest Conservancy in Victoria*, Government Printer, Melbourne, 1887, p. 8.

5 Alex Larkins, 'River and range: a story of a Victorian river valley', unpublished manuscript (Box 3075/1, MS 12377), La Trobe Collection, State Library of Victoria, p. 66.

6 *The Melbourne Historical Bottle Society Newsletter*, 1974–75, Rosanna.

7 Chris Alger, *A Puffing Billy Scrapbook: A Pictorial History 1940–88*, C R Alger for the Puffing Billy Preservation Society, Brunswick, 1988.

8 Interview with Frank Stamford, 25 February 1991; Frank Stamford, *Five to 500: the Light Railway Research Society of Australia's Twenty-five Year Book*, LRRSA, Melbourne, 1986, pp. 1–12; the LRRSA's journal, *Light Railways*, and its newsletter, *Light Railway News*.

9 *Light Railway News*, no. 70, June 1989, p. 4.

10 *Light Railway News*, no. 70, June 1989, p. 3.

11 *Light Railway News*, no. 64, June 1988, p. 3.

12 *Light Railway News*, no. 56, February 1987, p. 13; Stamford, *Five to 500*, p. 34.

13 Interview with Frank Stamford, 25 February 1991.

14 *Heritage News*, August 1990: 'To date there has been little systematic surveying of historical sites in forested areas', quoted (and contradicted) in *Light Railway News*, no. 79, December 1990, p. 3.

15 Interview with Frank Stamford, 25 February 1991.

16 A W Shillinglaw (Chief, Division of Forest Operations, Forests Commission of Victoria), 'Forests utilization', *Evidence Presented to the State Development Committee on Its Enquiry into the Utilization of Timber Resources in the Watersheds of the State*, Victoria government printer, Melbourne, 1959.

17 Shaun Lawlor, Darrin McKenzie and Jonathon Rofe, 'Forest harvesting: Survey of Christensen and Saxton's sawmill 1926–1937', industrial archaeology assignment, copy held in the Historic Places Branch, Department of Conservation and Environment, pp. 43–5.

18 Interview with David Ashton, 2 September 1991.

19 Land Conservation Council, *Wilderness: Special Investigation Descriptive Report*, LCC, Melbourne, 1990, p. 118.

20 Gunns Kilndried Timber Industries, Submission to Commission of Inquiry into the Lemonthyme and Southern Forests, Launceston, 1987 (submission by Brendan A Lyons), pp. 7–8.

21 *Light Railway News*, no. 70, June 1989, pp. 13–14; information from staff in the Powelltown office and Historic Places Branch of the Department of Conservation and Environment.

22 Peter Cabena, 'Submission to the LCC statewide assessment on post settlement history', unpublished report held in the Historic Places Branch, Department of Conservation and Environment. Part of this report was published by the Land Conservation Council in its *Statewide Assessment of Public Land Use*, LCC, Melbourne, 1988, pp. 150–9.

23 Theodore J. Karamanski, 'Logging, history, and the national forests: A case study of cultural resource management', *The Public Historian*, vol. 7, 2, Spring 1985, pp. 27–40.

24 Charles Fahey uses the term 'bureaucratic secrecy' in 'Review paper: keeping track of the world of foresters', in K J Frawley and N Semple (eds) *Australia's Ever Changing Forests*, Australian Defence Force Academy, Canberra, 1988, p. 460.

25 'Proposal to form an Australian Forest History Society', Appendix 1, in Frawley and Semple, pp. 527–8.

EPILOGUE

1 *Sydney Morning Herald*, 4, 6, 7, 8 January 1994; 'Failure to burn off blamed', *SMH*, 6 January 1994.

2 'Judge Stretton's reminiscences', *La Trobe Library Journal*, vol. 5, no. 17, April 1976, pp. 1–21; K Anderson, *Fossil in the Sandstone: The Recollecting Judge*, Spectrum Publications, Melbourne, 1986, pp. 124–30; E E Hewitt, *Judges Through the Years*, Hyland House, Melbourne, 1984, pp. 60–1; *The Australian Law Journal*, 30 September 1964, p. 183; *Law Institute Journal*, September 1967, pp. 360–1; *Australian Bar Gazette*, vol. 1, no. 4 , December 1964, vol. 2, no. 1, December 1966, p. 22; *Age* (Melb.), 18 May 1939, 4 August 1964, 17 May 1967; *Sun* (Melb.), 1 March 1939, 15 January 1953, 25 August 1964; *Argus*, 15 April 1937, 27 February 1952, 26 March 1952. I am grateful to Norman O'Bryan for locating several of these references.

3 Meredith Fletcher, 'Digging Up People for Coal: A History of Yallourn', PhD thesis, Department of History, Monash University, 1999, pp. 178–9.

4 L E B Stretton, *Report of the Royal Commission to Inquire into the Place of Origin and Causes of the Fires which commenced at Yallourn on the 14th day of February, 1944*, Government Printer, Melbourne, 1944, p. 3.

5 Fletcher, p. 213.

6 See Libby Robin, *Building a Forest Conscience: An Historical Portrait of the Natural Resources Conservation League of Victoria, 1944–1990*, NRCL, Melbourne, 1991.

7 L E B Stretton, 'Remember Black Friday!', *The Riverlander*, no. 66, January 1952, p. 3.

8 *Age*, 10 August 1937, p. 13; *Sydney Morning Herald*, 10 August 1937, p. 12.

9 *Argus*, 27 February 1952, 26 March 1952.

10 Evidence of James Francis Ezard, in 'Transcript of evidence', p. 26.

11 Stephen J Pyne, *Burning Bush*, Allen & Unwin, Sydney 1992, p. 312.

12 Hancock invoked Stretton in his Work-in-Prospect Seminar, History, Research School of Social Sciences, 7 March 1968, entitled 'Discovering Monaro – First Sight of a Seven Years' Task', Hancock Papers, National Library of Australia. See also his *South Australia's Lifeline* (A public lecture delivered at the University of Adelaide), Adelaide, 1983, pp. 16–17.

13 Peter M Attiwill, 'Ecological disturbance and the conservative management of eucalypt forests in Australia', *Forest Ecology and Management*, 63 (1994), pp. 301–46; David B Lindenmayer, 'Using environmental history and ecological evidence to appraise management regimes in forests', in Stephen Dovers (ed.), *Environmental History and Policy: Still Settling Australia*, Oxford

University Press, Melbourne, 2000, pp.
74–96; Andrew P Smith, 'Forest Policy:
Fostering environmental conflict in the
Australian timber industry', in Lunney (ed.),
Conservation of Australia's Forest Fauna,
pp. 301–14;

14 I have here drawn on my entry on
'Environmental History' in G Davison, J Hirst
and S Macintyre (eds), *The Oxford
Companion to Australian History*, Oxford
University Press, Melbourne, 1998.

15 Donald Worster, *The Wealth of Nature*,
Oxford University Press, New York, 1993,
pp. 24–5; Stephen J Gould, *Wonderful Life:
The Burgess Shale and the Nature of History*,
Penguin Books, London, 1991, pp. 51, 280–1;
Tom Griffiths, *Hunters and Collectors*,
Cambridge, Melbourne, 1996, chapter 1.

16 For example, J G D Clark, *Prehistoric Europe:
The Economic Basis*, Methuen & Co. Ltd,
London, 1952.

17 W L Thomas, Jr, (ed.) *Man's Role in
Changing the Face of the Earth*, University of
Chicago Press, Chicago, 1956. I am drawing
here on an unpublished paper by Richard
Grove, 'North American Innovation or
Imperial Legacy? Contesting and Re-assessing
the Roots and Agendas of Environmental
History, 1860–1996', presented at a
Colloquium on the Environment held at
the Research School of Social Sciences,
Australian National University, 14–15
February, 1996, kindly made available by
the author.

18 Grove, 'North American innovation or
imperial legacy?', and *Green Imperialism:
Colonial Expansion, Tropical Island Edens
and the Origins of Environmentalism,
1600–1860*, Cambridge University Press,
Cambridge, 1995.

19 Michael Williams, 'The relations
of environmental history and historical
geography', *Journal of Historical Geography*,
vol. 20, no. 1, 1994, pp. 3–21.

20 Greg Dening, 'A Poetic for Histories:
Transformations that Present the Past', in
Aletta Biersack (ed.), *Clio in Oceania*,
Washington, Smithsonian Institution, 1990,
pp. 347–80; Griffiths, *Hunters and Collectors*,
pp. 210–11.

21 Alfred W Crosby, 'The past and present of
environmental history', *American Historical

Review*, vol. 100, no. 4, October 1995,
pp. 1177–89, at p. 1181.

22 For a commentary on the concern with the
agency of nature, see David Demeritt, 'The
nature of metaphors in cultural geography and
environmental history', *Progress in Human
Geography*, vol. 18, no. 2, June 1994,
pp. 163–85.

23 Roderick Nash, *The Rights of Nature: A
History of Environmental Ethics*, University of
Wisconsin Press, Wisconsin, 1989.

24 Roderick Nash, 'American environmental
history: a new teaching frontier', *Pacific
Historical Review*, vol. 41, 1972, pp. 362–71.
See also Donald Worster, 'Doing
environmental history', in Worster (ed.), *The
Ends of the Earth: Perspectives on modern
environmental history*, Cambridge University
Press, Cambridge, 1988, pp. 289–307.

25 Stephen J Pyne, preface to new edition of
Burning Bush: A Fire History of Australia,
University of Washington Press, Seattle, 1998.

26 Donald Worster, Alfred W Crosby, Richard
White, Carolyn Merchant, William Cronon
and Stephen J Pyne, 'A round table:
Environmental history', in *Journal of
American History*, vol. 76, no. 4, March
1990, pp. 1087–147.

27 William Cronon, 'The uses of environmental
history', *Environmental History Review*,
vol. 17, no. 3, Fall 1993, pp. 1–22, at p. 13,
and Donald Worster, 'The two cultures
revisited: Environmental history and the
environmental sciences', *Environment and
History*, vol. 2, no. 1, February 1996,
pp. 3–14. See Cronon's *Nature's Metropolis:
Chicago and the Great West*, W W Norton &
Co., New York, 1991, and Worster's,
'Transformations of the Earth: Toward an
agroecological perspective in history', in
Worster *et al*, 'A Round Table', pp. 1087–106.

28 Worster, 'Doing environmental history', p. 290.

29 See, for example, Crosby, 'The past and
present of environmental history', and
Worster, *The Wealth of Nature*, p. 45.

30 John MacKenzie, 'Empire and the ecological
apocalypse: The historiography of the imperial
environment', in Tom Griffiths and Libby
Robin (eds), *Ecology and Empire:
Environmental History of Settler Societies*,
Keele University Press, Edinburgh, 1997,
pp. 215–28.

31 Donald Worster, 'The Vulnerable Earth: Toward a planetary history', in Worster (ed.), *The Ends of the Earth*, pp. 3–20.

32 Libby Robin, *Defending the Little Desert: The Rise of Ecological Consciousness in Australia*, Melbourne University Press, Melbourne, 1998; Richard Grove, *Green Imperialism*. On environmental history and environmentalism, see J M Powell, 'Strangers and lovers: Disputing the legacy of environmental history', in Livio Dobrez (ed.), *Identifying Australia in Postmodern Times*, Bibliotech, Canberra, 1994, pp. 87–103.

33 Robin, *Defending the Little Desert*, especially chapters 4 and 8.

34 Donald Worster, *The Wealth of Nature*, Oxford University Press, New York, 1993, p. 158.

35 Worster, *The Wealth of Nature*, p. 164.

36 Worster, *The Wealth of Nature*, p. 169.

37 Stephen J Pyne, 'Smokechasing: The Search for a Usable Place', unpublished paper kindly made available by the author at http://www.public.asu.edu/~spyne/Smokechasing.htm.

38 Worster, 'History as natural history' and 'The ecology of order and chaos', in *The Wealth of Nature*, pp. 30–44 and 156–70.

39 Richard White, 'Environmental history, ecology, and meaning', in Worster *et al*, 'A Round Table', p. 1115.

40 Stephen J Pyne, *Burning Bush*, (1998) preface.

41 J M Powell, 'Historical geography and environmental history: an Australian interface', *Journal of Historical Geography*, vol. 22, no. 3, 1996, pp. 253–73, at p. 255. See also J M Powell, 'Strangers and lovers' (see note 32), 'Marginal notes? Recent works in Australian environmental history', *Australian Geographer*, vol. 27, no. 2, 1996, pp. 271–8, and *Disputing Dominion: Environmental sensibilities, historical consciousness and academic discourse in Australia*, Working Paper No. 36, Department of Geography and Environmental Science, Monash University, Melbourne, 1995.

42 Pyne, Preface, *Burning Bush*, 1998.

43 William Cronon, 'A place for stories: Nature, history, and narrative', *Journal of American History*, vol. 78, 1992, pp. 1347–76, at p. 1375.

44 See Tom Griffiths, 'The writing of *A Million Wild Acres*', in John Dargavel, Di Hart and Brenda Libbis (ed.) *The Perfumed Pineries*, Australian Forest History Society, Canberra, 2001.

45 I have not discussed here, for example, the significant work of W G Hoskins and J B Jackson, or the influence of Fernand Braudel, because I am focusing on the distinctive character of environmental history since 1970, but see Tom Griffiths and Tim Bonyhady, 'Landscape and language', in *Words for Country: Landscape and Language in Australia*, University of NSW Press, Sydney, 2001, and Griffiths, 'Travelling in deep time: *La Longue Durée* in Australian History', *Australian Humanities Review*, 2000.

INDEX